Counting Backwards

Counting Backwards

A DOCTOR'S NOTES ON ANESTHESIA

Henry Jay Przybylo, MD

W. W. NORTON & COMPANY

Independent Publishers Since 1923

NEW YORK LONDON

Counting Backwards is a work of nonfiction. Patient names have been changed, and potentially identifying details and distinguishing facts and characteristics have been altered or amalgamated. No individual who appears in these pages should be construed to refer to or identify any single patient

The views and opinions expressed in this book are solely those of the author. They do not reflect the views or opinions of any organization or institution with which the author is or has been affiliated, or of anyone else employed by or affiliated with such organizations or institutions.

Copyright © 2018 by Henry Jay Przybylo, MD

All rights reserved
Printed in the United States of America
First published as a Norton paperback 2018

For information about permission to reproduce selections from this book, write to Permissions, W. W. Norton & Company, Inc., 500 Fifth Avenue, New York, NY 10110

For information about special discounts for bulk purchases, please contact W. W. Norton Special Sales at specialsales@wwnorton.com or 800-233-4830

Manufacturing by LSC Communications Harrisonburg
Book design by Barbara Bachman
Production manager: Beth Steidle

LIBRARY OF CONGRESS CATALOGING-IN-PUBLICATION DATA

Names: Przybylo, Henry Jay, author.
Title: Counting backwards : a doctor's notes on anesthesia /
Henry Jay Przybylo.
Description: First edition. | New York : W.W. Norton & Company, [2018]
| Includes bibliographical references.
Identifiers: LCCN 2017018049 | ISBN 9780393254433 (hardcover)
Subjects: | MESH: Anesthesiologists | Anesthesia | Illinois |
Personal Narratives
Classification: LCC RD80.62 | NLM WO 221 | DDC 617.9/6092—dc23
LC record available at https://lccn.loc.gov/2017018049

ISBN 978-0-393-35642-7 pbk.

W. W. Norton & Company, Inc.
500 Fifth Avenue, New York, N.Y. 10110
www.wwnorton.com

W. W. Norton & Company Ltd.
15 Carlisle Street, London W1D 3BS

1 2 3 4 5 6 7 8 9 0

To my wife who taught me
how to love and how to live.
Who encouraged my growth and
was my every inspiration.

And who left me far too soon.

Sandy Przybylo d. October 19, 2015

Contents

Introduction   ix

CHAPTER 1   A Deep Sleep   1

CHAPTER 2   Command Center   10

CHAPTER 3   The Five A's   31

CHAPTER 4   Railroad Tracks   47

CHAPTER 5   Fear of the Mask   57

CHAPTER 6   Nothing by Mouth   81

CHAPTER 7   Heartbeats   97

CHAPTER 8   A Most Unusual Patient   115

CHAPTER 9   Errors Everlasting   132

CHAPTER 10   In Wait   149

CHAPTER 11   Paper Cranes   161

CHAPTER 12   A Brain Trapped in a Box   181

CHAPTER 13   See One, Do One, Teach One   191

CHAPTER 14   Reentry   211

CHAPTER 15   Safe Travels   219

Acknowledgments   229

A Note on Sources   231

Introduction

———

AM AN ANESTHESIOLOGIST. I ERASE CONSCIOUSNESS, deny memories, steal time, immobilize the body; I alter heart rate, blood pressure, and breathing. And then I reverse these effects. I eliminate pain during a procedure, and prevent it afterwards. I care for sick people and I have saved lives, but it's rare that I'm the actual healer. As an anesthesiologist, I do nearly all of my hands-on work behind automatic double doors, sequestered, allowing surgeons to cut, gastroenterologists to probe, cardiologists to stick. The patients I care for place their faith in me, but we've usually been introduced only a few minutes before, and they rarely remember my name after their surgery or procedure is completed.

I put people into a coma, and the medications I administer cause paralysis. Yet only a handful of times each year do patients or family members ask how anesthesia actually works. The truth is there is much about anesthesia that even modern science can't yet explain.

But I know, as sure as the sun rises in the morning, that when I add a gas to the inhaled breaths, loss of consciousness follows; and when I remove the gas, awareness returns. This is a harrowing responsibility, and one I never take for granted.

Forty million people in the United States undergo anesthesia every year. It is the most frequently performed medical procedure that entails risk to the patient. The anesthesiologist is ubiquitous but largely invisible. Before the slash of the scalpel, the insensibility to pain is taken for granted. More people, up to a hundred million per year, seek relief from pain, both acute and chronic. Pain is the most common human health issue.

As an anesthesiologist of more than thirty years' experience, working in a come-one-come-all practice within a large university health care system, I have administered anesthesia more than thirty thousand times during my career. To newborn babies, to kids, to adults in the prime of life, to centenarians. To patients dealing with the most benign conditions (removing a skin mole or placing infection-clearing tubes in eardrums) and to those facing potentially fatal ones (clipping a cerebral artery aneurysm). I specialize in pediatric anesthesiology, caring for about a thousand children in an average year—from one-and-a-half-pound micropreemies with skin so new that the tissues and bones beneath are clearly visible, to massively obese teenagers.

Anesthesiologists don't keep regular hours; our care and expertise are called upon at all times, from midday elective colonoscopies to middle-of-the-night emergency trauma. There's always an on-call anesthesiologist ready to respond. Day in, day out, most of the waking hours of my life are spent in that cloistered place behind the automatic double doors.

Anesthesiology allows little of what was learned in medical school to be forgotten. Perhaps no other specialty remains as expansive or inclusive, covering all the basic sciences (anatomy, pathology, physiology, pharmacology) and all fields of clinical medicine (internal medicine, surgery, pediatrics, obstetrics, and even psychiatry), and interacting with every other conceivable specialty. On any given day, any page of the physiology, pathology, or pharmacy texts may need to be scoured for a reference. From the time we meet in the preanesthesia area until the point that, after the procedure is over, I'm assured my patient has returned to a state of comfort and is prepared to reunite with loved ones, I am the primary care physician. During my anesthesia care, I become the internist, the ob-gyn, the pediatrician. The child scheduled to have a skin mole removed might have a failing heart; the woman whose brain aneurysm has burst might also suffer the pain and deformities of rheumatoid arthritis. When things take a turn for the worse during a procedure, when the blood loss climbs or the heart rhythm goes awry, it's left to the anesthesiologist to make life right.

Over 170 years ago, inhaling a gas was shown to render a person senseless and thereby allow invasive medical procedures. Health care exploded. The magnitude of this medical discovery remains obvious today. The number of patients receiving anesthesia grows every year. The renowned *New England Journal of Medicine* recently polled its readers to select the most important article the magazine had published in its distinguished history. The "resounding favorite" its readers settled on was the *NEJM*'s 1846 article by Henry Jacob Bigelow about the demonstration showing that inhaling ether allows for pain-free surgery. ("Anesthesia" as a term had not yet been coined.) This article, published just a few months after the historic demonstration at Mass General's Ether Dome in Boston, bested every advance made since then, including the introduction of antisepsis, of X-ray imaging, and of antibiotics.

More than a century and a half later, I am unable to provide an answer when patients and families ask the most basic question of my specialty: how the gas I use anesthetizes. Despite decades of research, its mechanism of action remains a mystery. I must have faith in my anesthesia gas. It's an irony of our work that patients and their loved ones place faith in the anesthesiologist, who in turn places faith in the gas. In many ways, I'm a faith healer.

I routinely ask patients to count backwards as the anesthesia medications are introduced. Counting backwards in this way is a time-honored anesthesiology tradition. When rapid-acting barbiturates were introduced half a century ago, making it possible to induce loss of consciousness in a matter of seconds, an anesthesiologist likely asked a patient to start at 100 and count back, curious to determine the speed of induction. *100 . . . 99 . . . 98 . . .*

The practice stuck.

In my experience, patients never make it out of the 90s.

Counting Backwards

CHAPTER 1

A Deep Sleep

AMANDA NEEDED A DEEP SLEEP. A SLEEP LIKE no other that she had experienced in her five years of life. Breathing was difficult, her nose always congested. When she did sleep, she snored and her nose frequently ran. A clump of tissue with the consistency of jam, the adenoids, blocked the breathing path through her nose. Her surgeon needed to pass large instruments through her mouth, beyond her tongue, and past her tonsils to scrape or burn—the exact technique the surgeon's choice—the redundant adenoid tissue to open the path of air through her nose. The procedure required Amanda to remain still, open her mouth, and keep it open; to let the surgeon pass surgical instruments deep into her mouth; to not cough or gag; and to not scream or cry when the knife sliced the adenoids out. To make all this possible, Amanda needed the deep sleep of anesthesia.

Amanda knelt on the gurney, her toes peeking out from under her bottom. She leaned forward, elbows on the mattress, contentedly coloring the paper in front of her, unaware of what was to come. I don't think she even noticed me when I entered her pre-anesthesia space. Her parents, standing to the side of the gurney, appeared trapped in tight quarters and couldn't disguise the fear on their faces. How would I place their daughter under the influence of my gas, and would it be safe?

"In the thousands of cases in my career, my anesthetic has never failed. It's one hundred percent effective," I said. The look in their eyes changed slightly, but I didn't have their complete confidence yet.

The health care buzzword of the day is "transparency." Describe every treatment option, all the benefits and possible complications, then let the patient, or in Amanda's case her parents, make the decision. I've signed consents for procedures for others—my kids and wife. No greater stress in life exists than making decisions—decisions with possible lifelong implications— for another person, even if you brought that person into this world. That stress increases as the procedure approaches, and the more I say, the less my patients and their families hear. Patients come to me for my insight and expertise. They want to know what I recommend. They want to know what I would do for myself or my family. Many don't want to make a decision, especially one they are not well versed in. Amanda's parents didn't understand anesthesia or my intended process. I explained my plan as plainly as possible.

"It's as simple as this. From the time I pass through those doors," I said, gesturing to the sliding glass doors of the prep room, "until the time she's asleep will take less than two minutes. The mask goes on and it takes only eight to ten breaths."

Amanda's parents probed for assurance and confidence.

"If there's a problem, it's mine. You come to me. I'm the one responsible for anything that might go wrong. And I don't like problems, so there won't be any."

I mentioned that I would love to etch a guarantee in granite, but I can't. I added: "In all the thousands of my cases, a healthy patient goes in, a healthy patient comes out."

A look of amazement swept over their faces.

"At the end of the surgery, I turn my gas off, Amanda breathes fresh air, and the anesthesia wears off. Technically, she's awake when she leaves the room, but recognition takes a few minutes. She'll be in recovery when she understands where she is. As soon as that happens, you'll rejoin her."

In all, from the time I first meet patients, families, and loved ones until the trip to the anesthetizing area begins takes about three minutes. In that time, I need to earn their trust and have them place their faith in my care.

Soon, Amanda would be in the procedure room breathing a combination of gases on her way to entering the state of anesthesia.

A MAGIC PILL THAT RELAXES, soothes, comforts, prevents all pain, staves off bad dreams and thoughts, and provides a cooperative patient during medical procedures. Such a magic pill is the pharmaceutical industry's ultimate fantasy. In reality, it's already been found, and it's not a pill and it's not natural. It's a gas that's synthesized.

A volatile is a fluid that in air at room temperature would rather exist as a gas. Volatiles are part of everyday life. They are components of cleaning fluids, bleach, paint, nail polish remover, and, most important for me, my anesthesia gases. The term "volatile anesthesia" means breathing a gas to induce the anesthetic state. Ether is one such volatile that loves to become a gas. Its magical benefit to health care—painless surgery—was first shown in the 1840s, but its history goes much farther back in time.

The Muslim cleric and alchemist Jabir Ibn Hayyan, who lived in Persia in the eighth century, likely came close to synthesizing ether. He is known to have used the components needed to synthesize it, including sul-fur, which is necessary to create the reaction with alcohol. But it remains speculation that Jabir actually managed to produce ether. (Jabir was a remarkable man regardless of his role in synthesizing ether. As a noted philosopher, geographer, and linguist, he is said to have written some three thousand books. The Latin form of Jabir is "Geber," and his prolific and wide-ranging writing is considered by experts in etymology to have inspired the word "gibberish.")

In 1540, Valerius Cordus, a German physician, botanist, and alchemist whose life was only a flash—he was twenty-nine when he died—combined fortified wine and sulfuric acid to form what he called *oleum dulce vitrioli*, the delicious Renaissance Latin term translated as "sweet oil of vitriol." Its medicinal properties were noted and came to be of even greater interest when Cordus's contemporary Paracelsus, the Swiss-German polymath, found that ether caused chickens to sleep. Paracelsus is believed to have tested ether to treat seizures, the result unclear. He might have gone on to discover the painless state created by the sweet oil of vitriol, allowing for surgical invasions of the body, but he, too, died prematurely and mysteriously.

Another two centuries passed before the German-born chemist August Sigmund Frobenius published an article in 1729 describing the method of synthesizing *oleum dolce vitrioli* and calling it "ether," from the Greek base word meaning "to alight" or "to blaze" (ether is very flammable). Ether also connotes the upper air, which is fitting because it is a volatile that readily changes from liquid to gas.

Near this time, pneumatic medicine—inhaling gases as a means of therapy—flourished, leading to the discovery of the euphoria one can attain by inhaling the fumes of ether. Medical students, then as today, searched for new means to distract themselves from the pyretic intensity of their work and found that "ether frolics" provided that relief.

Medical cures of the time weren't founded in science and included treating asthma with dried, ground toad; holding a live puppy to the stomach to deal with bowel obstructions; applying leeches to bleed bad humors; and using dog feces as a remedy for sore throat. But as alchemy—the belief of transforming one compound into another, such as lead into gold—evolved into chemistry, the individual gases composing air (oxygen, nitrogen, and carbon dioxide) were isolated. Joseph Priestley, a chemist and noted grammarian—he wrote *The Rudiments of English Grammar*—produced nitrous oxide for the first time in 1772. By 1800, the chemist Humphry Davy had noted that this colorless and odorless gas caused a state of euphoria and might even be used to prevent pain during surgery. He didn't pursue that purpose. By the turn of the century, nitrous oxide demonstrations for comic relief had become a business.

In the 1830s, Samuel Colt, under the name of the "Celebrated Dr. Coult of New York, London and Calcutta," demonstrated the effects of nitrous oxide, then encouraged audience participation, for twenty-five cents per man. He advertised with a poster declaring "A Grand Exhibition . . . Laughing Gas. Laugh, Sing, Dance, Speak or Fight" and used his profits to develop the Colt revolver. In New York a few years later, P. T. Barnum opened Barnum's American Museum, where visitors could test laughing gas. Traveling shows charged "gentlemen of the first respectability only," although one poster included a drawing of a woman, for the chance to inhale this gas that offered short-lasting, non-alcohol-altered sensory impairment, with no hangover.

IT WAS 1839 IN RURAL GEORGIA, and a slave was forced to inhale the vapor of ether as onlookers sought to have him dance while impaired. The prank ran sour when the slave boy—whose name was never recorded—passed out for an extended time. Frightened, the party-goers summoned a doctor, who observed the boy until the ether wore off with no apparent ill effect. The intention of such forced ether frolics appears to have been not to lose consciousness, but to stumble uncontrollably for the amusement of the observers. Word of the misadventure spread to other doctors in the area.

Near the same time, a Georgia doctor trained in Pennsylvania, Crawford Long, brought ether frolics to his own community. Long is thought to have known of the 1839 incident, which set the foundation for what followed. In Long's own words: "In the month of Dec. 1841, or in Jan. 1842, the subject of the inhalation of nitrous oxide gas was introduced in a company of young men assembled at night in the village of Jefferson, Ga., and the party requested me to prepare them some. I informed them I had not the requisite apparatus for preparing or preserving the gas, but that I had an article (sul. ether) which would produce equally exhilarating effects and was as safe." After the effects of the inhaled ether wore off, Long noticed new scrapes and bruises that he couldn't explain. He observed others sustaining cuts and bruises without reaction while under the influence of the gas.

An acquaintance of Long, a young man named James M. Venable of Jackson County, Georgia, found himself in the right place at the right time and in the right company. A few months shy of his twentieth birthday, he sought out Long's advice regarding a lump on his neck. Long advised that the lump be surgically removed. Venable dreaded pain, but Long assured him that the lump could be removed without it. On March 30, 1842,

Long folded a towel, saturated it with ether, covered Venable's mouth and nose, and instructed him to breathe. Minutes later, Venable emerged from unconsciousness proclaiming he felt no discomfort, as the resected mass sat in a surgical pan. It was the first known use of inhaled ether to enable safe, painless surgery.

Long did not document the event at the time. He did, however, record the two-dollar charge for ether—the first known bill for providing painless surgery. Long, and his family and colleagues, later offered plausible excuses for his failure to publish and establish his claim. He lived twenty miles from the nearest published paper and several times farther from the closest medical college. He believed he would need to report on a series of cases, not just a single patient, to merit publication. And at just twenty-nine, he doubted that his veracity would be accepted by elder physicians. A deep sleep removing all senses from a person was also perceived as sacrilegious at the time. An unnamed clergyman later decreed: "Anesthesia is a decoy of Satan . . . rob God of the deep earnest cries which arise in a time of trouble." In rural Georgia, Long might just have been right to hold his secret near and dear to himself.

In Connecticut in 1845, a dentist named Horace Wells watched a Barnum protégé demonstrate laughing gas. The performance convinced him that he'd found the means to pain-free dentistry. The day after, Wells inhaled nitrous oxide himself as a colleague pulled one of his teeth. Wells hastily arranged a public exhibition of the use and effect of inhaled nitrous oxide. Boston was selected as the site for his demonstration in December 1844. In a hall lost to history, Wells administered nitrous oxide to a medical student for the extraction of a tooth. Nevertheless, the student cried out in pain during the procedure. The idea of potency was not yet known; nitrous oxide lacks sufficient strength to create pain-free conditions for more invasive procedures. Wells was showered with ridicule for his failure and labeled a tarnished idol. He gave up dentistry, and the ensuing descent led to his 1848 addiction to another inhaled gas found to alter the senses, chloroform. Wells committed suicide in a New York prison after a deranged attack in which he poured acid on two women.

William Morton, a partner in dental practice with Wells, understood the significance of inhaling a gas to relieve procedural pain. With Wells's failed demonstration, their partnership dissolved and Morton turned to a previous teacher, Charles T. Jackson, who advised him to try ether. After experimenting with ether—there is some debate as to how many patients he experimented on—Morton arranged for a public demonstration in 1846, also in Boston, of painless surgery to remove a jaw mass. But before his show, knowing the magnitude of his discovery, he first visited the patent office. His application described both the administration of a gas to produce insensibility to pain and the device used to present the gas to the patient. Morton had designed a glass vessel that contained a sponge soaked with the fluid ether and a wooden mouthpiece that extended from the vessel to the patient. Morton went to great lengths to ensure that his discovery was accurately documented, both in the newspapers and journals of the time, as well as legally.

Morton understood the advantage of nitrous oxide (that it was odorless) and the disadvantage (that it was too weak). He accepted the distinct and strong smell of ether because, unlike nitrous oxide, it rendered the patient unconscious. Morton not only tried to contain the odor in a flask so that others around might not recognize it, but he also disguised it by adding citrus oil to the ether.

On October 16, 1846, Morton conducted his demonstration on a patient named Edward Gilbert Abbott in the operating room of the Massachusetts General Hospital. When the patient breathed the gas, Morton announced to the surgeon: "Sir, your patient is ready." Surgery commenced, and the patient did not react to the slash of the scalpel. When it was over, the surgeon, James Warren, turned to the audience and declared: "Gentlemen, this is no humbug." The operating room used for the demonstration gained a new name that remains to this day, the Ether Dome, and it is now a declared historic site. The published paper announcing the arrival of painless surgery was not written by Morton or Warren, but by another surgeon observing the procedure, Henry Bigelow, who had helped arrange the demonstration.

Morton's patent for what he called "Letheon" won approval. Morton chose the name from the River Lethe in Greek mythology; drinking the water of the river caused loss of memory. (The name seems to me too close to "lethal," from the Latin *letum*, meaning "death," but it didn't matter.) The size and configuration of the Ether Dome placed many observers within reach of the patient and close enough to smell the gas. Those present at the demonstration saw through the deception and recognized the odor of ether, a readily available chemical. Further mention of the citrus oil disguise was dropped.

News of painless surgery after inhaling ether spread across the world with stunning speed. In an era before the Pony Express or the telegraph, it took only a few months following the Ether Dome demonstration for articles describing the discovery to show up in newspapers from Hawaii to Paris. The news traveled so fast and the use of ether became so frequent that Francis Plomley soon wrote an article in the British medical journal *Lancet* describing the stages of anesthesia—from too little inhaled gas, which left the patient giddily impaired but not yet ready for the scalpel; to excitation; to just right, allowing surgical anesthesia.

A month later, the term "anesthesia," Greek for "without sensibility," was coined in a letter to Morton by Oliver Wendell Holmes Sr.—the Boston poet, physician, and professor—and it stuck. "Letheon," the name for ether listed by Morton on his patent application, didn't.

AMANDA'S MOTHER, UNAWARE OF anesthesia's history and unconvinced by the description of my process, remained dubious that I could induce a chemical coma in Amanda in a minute or less. She insisted on accompanying her daughter to the OR.

The gas I would use to anesthetize Amanda, sevoflurane, differs remarkably little from the ether used in 1846. Today's gas has the same chemical backbone: four C's (carbon) anchored by an O (oxygen). Instead of ten H's (hydrogen), seven are replaced by jumping two letters to the left in the alphabet, to F (fluorine). In the 160 intervening years between ether and sevoflurane, many different gases were introduced, all with one or another undesirable trait and left to remain in historical context only. Sevoflurane is nonflammable, much less odorous than ether, and—most important—dramatically decreases the time until loss of consciousness. Two other volatile gases are commercially available—desflurane and isoflurane—but the properties of sevoflurane led to its popularity. (It appears we've reached the end of the line with volatile anesthetic gases; the ones in use today were all discovered decades ago and no new agents are forthcoming.)

Nitrous oxide, although not potent enough to use alone, is added because it provides a boost to the loss of sensation, and it does so without an unpleasant odor. I planned to start Amanda's anesthetic experience by having her breathe nitrous oxide at fifty percent of the inhaled gas. Amanda would be shown a mask and asked to choose a scent—bubblegum, cherry, strawberry, or orange—which would be rubbed on the inside of the mask to cover the scent of the sevoflurane she would breathe. And she would have little to no recall for the odor of induction of anesthesia if I could convince her to breathe the nitrous oxide for thirty or so seconds before I added the potent sevoflurane.

The state of anesthesia is far different from sleep, but every day I find myself commenting to my patients and families on the process of "going to sleep." During the induction of anesthesia, I advise patients: "Pick out a good dream. Do you have a happy place? Go there." But there are no dreams under the influence of anesthesia.

Since I am not especially religious, it was decades after entering medicine before I suddenly saw a connection between the Bible and anesthesia. Genesis 2:21: "And Yahweh caused a deep sleep to fall upon Adam, and He took one of his ribs, and closed up the flesh." Did Adam receive anesthesia for his rib resection?

To calmly anesthetize a child requires a skill set different from that needed to anesthetize an adult. An adult in the holding area will volunteer an arm, remain cooperative as the tourniquet tightens, and not flinch as the needle pierces the skin and enters a vein. Then, after a relaxing injection, it's off to the procedure room for the countdown in reverse. I make it a little more challenging. As the intravenous anesthetic agent flows in, I suggest to start at 100 and count backward by 7's. Making it to 93 is tough.

With a child in the holding room, just mention a shot and a struggle ensues.

I spoke to Amanda on the trip to the OR. Amanda's mother followed. I pointed out the bees, butterflies, and birds painted on the walls and talked about her favorite colors. In the operating room, I continued to talk about breathing bubblegum, the scent she had chosen for the mask. As long as Amanda could remain calm, having her take several breaths of the gas that had failed Wells, nitrous oxide, would likely prevent her from remembering the odor of the potent anesthesia gas. Then we talked about a pretend trip to the zoo and smelling piggies.

"How many piggies do you smell?"

"Five piggies," she said as she drifted into a state of anesthesia quickly and quietly.

I turned to her mother and said: "She's asleep."

CHAPTER 2

Command Center

WASN'T BORN A CONTROL FREAK. I GREW INTO ONE, and slowly at that. I was the typical teen, lackadaisical and rudderless, leaving a sloppy trail of uncompleted projects behind me. But today, within the nest of my anesthesia practice, I'm all about control. And I am always searching for ways to improve my care.

Many years ago, when I was just a year or so into my anesthesia training, a woman in her thirties lay on the OR table in front of me. Otherwise quite healthy, she possessed a lump in her neck, a thyroid nodule, that raised a ruckus by producing excess hormones—a condition termed thyrotoxicosis. She suffered from nervousness, fast heart rates, heat intolerance, and excessive perspiration. Her surgeon, Fast Eddie—the nickname he earned for his surgical prowess—approached the status of magician for the way he unzipped the thyroid gland from its home in the neck. In a teaching hospital, commonly a university-affiliated medical center, surgeons aren't necessarily slaves to the clock. Residents, both in anesthesia and surgery, consume time. Slow-as-molasses surgeons who would flail about trying to succeed in a time-driven private practice find acceptance in these centers. So, Fast Eddie was an anomaly. That malfunctioning thyroid dropped into the pan in spectacular time.

Contemplating the end of my anesthesia care as the surgery neared completion, I pushed the contents of the syringe that I *thought* contained the drug intended to reverse the muscle paralysis, which I had injected at the start of the case. Then I noticed the wrong label on the syringe in my hands. My heart skipped a few beats. Syringes are wrapped with an identifying color-coded tape for the various types of drugs I use. The orange on the syringe I had injected indicated that the drug I'd just pushed wasn't the reversal drug I had intended to use. It was relaxant, the paralyzing drug that I'd used at the start of the anesthesia. Before I could react and stop the IV, the drug flowed into my patient's bloodstream. The unintended drug was safe, but it would keep my patient unable to breathe on her own for an extra hour before wearing off.

I had just executed the classic syringe swap, and I was the fool.

"Doctor Ed," I said. "I just made a mistake that's going to cost you some time."

The surgeon stopped suturing the neck incision and looked up at me.

"I just reversed with relaxant."

Fast Eddie understood. He raised his eyebrows, smirked visibly even through his mask, dropped his head back to the surgical field, and continued suturing. The worst of his response was that he didn't say a word.

Now, as a complication, this wasn't life threatening or even dangerous during this stage of my care. It added unnecessary time to the procedure and required me to breathe for my patient for that extra hour until the effect of the paralyzing drug wore off, but she didn't suffer from my error. Still, to me my mistake was damning and painful. I had blundered. Letting down any proceduralist is beyond embarrassment; disappointing a superb clinician, someone practiced at perfection, is haunting. The enduring image of Fast Eddie's expression found a permanent place in my memory vault. I recall the room, the time of day, and the surgeon. His look of disappointment is a frequent reminder not to err again.

After that I stepped back and contemplated my setup at length. This blunder ignited my transition to control freak. Previously, when instructed, I set up the room and learned by trial and error. I had no cheat sheet, no set of rules, no deep, thoughtful process. But humbled by error, I began a continuing education in process, quality, and critical-incident theory that continues to this day.

In my experience, a critical incident—in my anesthesia world, a "complication"—results from many smaller mistakes intersecting at a common point. For example, an anesthesia technician stocks the room, and a resident prepares for the case without confirming the technician's work. A visitor to the OR, maybe a student, then trips and falls across the anesthesia circuit, disconnecting the patient from the ventilator. The resident evaluates the situation and concludes that the emergency resuscitation bag is the only manner to quickly restore breathing for the patient, but there is no resuscitation bag in the bottom drawer of the anesthesia cart. If the cart had been properly stocked by the technician, if the anesthesia resident had checked the cart, if the student had been more observant, or if the resident had guarded the anesthesia circuit from becoming dislodged—if any single miscue had not happened—the complication would not have occurred.

IT MAY SEEM A BIT of a reach to connect the contributions of Henry Ford to my medical career. He was not an inventor or scientist, but he was undeniably a great innovator, developing the concept of the production line as applied to building cars and setting the standard for the production of nearly everything today. In truth, the inventor of the production line was Ransom Olds, who patented the process of using an assembly line to mass-produce cars in 1901. A decade later, Ford expanded the concept to vastly increase production. His goal of supplying cars for the multitudes included a secondary gain of efficiency.

Quality measures freedom from error. A logical step after developing the production line was to measure how well it worked. Aside from the number of units built in a given amount of time, another measure of success is how well each unit is built. W. Edwards Deming proposed the first measures of quality in the post–World War II era, again applied to car manufacturing, as he developed fourteen key principles to improve quality. He used statistical methods to analyze for errors.

Analyzing the quality of a physician's care of the human body using the standards of car manufacturing and Deming's statistics was long considered impossible. The human body was thought to be far too complicated a system, composed of seven octillion cells functioning simultaneously, with any single cell able to fail. Cell functions seemed difficult to evaluate, and the reporting of symptoms was riddled with subjectivity. Relieving back pain, for example, might be considered a success at one medical center but a failure at another. In the mid-1980s, a federally funded RAND report published an equation for quality of health that included a variable for "full emotional well-being"—try putting a number on that—and concluded that "quality assurance is unlikely to grow in prominence."

But around that time, Jeffrey B. Cooper, a pioneering engineer who researched patient safety, began showing that human error led to anesthesia mishaps and that "critical incident" theory could be applied to improve quality. The critical-incident concept entered anesthesiology with the realization that one big error culminated from the sum of numerous small mistakes. For example, if a patient doesn't speak English, questions during a preoperative evaluation with the patient's internist might be translated by a relative and the note then entered not in the medical record, but in a letter. At the hospital, the anesthesiologist's pre-anesthesia evaluation could take place with a different translator. The evaluations might differ, and a previous anesthesia complication discussed with the internist might not be elicited by the anesthesiologist. If the internist's note is not seen, a complication could occur.

Invoking efficiency standards to improve health care treads perilously close on the sanctity of life. It's hard to accept that humans are merely a biological machine, although the most complex one in our world. Medicine might seem beyond applying the principles of a production line. But both manufacturing and medical procedures benefit from employing the same process, the same way, every time.

The solution to my syringe swap was intuitively simple: specify a distinct place for every one of the medications I use during the anesthesia process. For every patient, for every case, I draw into and label syringes the same way every time. No exceptions. I developed a set process for syringe placement, and the reversal agent is nowhere near the relaxant drug.

ONCE I STAND AT THE HEAD of the procedure table, it's time for business. Searching for drugs or equipment in a moment of need indicates below-standard preparation. I must be ready for any potential issues and pitfalls specific to the patient's illness, condition, and intended procedure. I work under my own corollary of Murphy's law: *If you have it, you won't need it.* The addendum: *Act expeditiously.* The basic needs must be within an arm's reach and not concealed in clutter. The anticipated is one step away; the potential, another step beyond. During anesthesia, efficiency correlates with economy of thoughts and movements.

The obsessive-compulsive in me materializes. With the diligence of a priest preparing for a service, altar and all, I prepare my space, equipment, and drug, clearing my mind, time, and energy to care for my patient.

I find nothing wrong with a preprocedure invocation. A short prayer or a moment of silence might seem an appropriate way to approach caring for every patient; after all, there is a good deal of faith in the practice of anesthesiology. The truth is, except in rare cases, I don't pray—by which I mean I don't rely on prayer. Or luck. Relying on luck in the procedure room is a bad choice.

Instead, I rely on a simple thought to gain focus, to be spot on. In the words of the screenwriter John Hughes in his movie *The Great Outdoors*: "Our Lady of Victory, pray for us." "Game on" would work just as well. Invoking hope is left to those in waiting. I also envision: "Dancing fingers." "Be the vein." "Own the airway." "See the whole picture." "Observe the panoramic everything." How quickly can the needles I place find the mark? How close to the target heart rate and blood pressure can I keep the patient? I am not searching for luck. I am striving for pinpoint skill.

Entering the procedure room is like entering a place of worship. The space is instantly recognizable, and its intended use clear. The procedure table sits in the room proper like an altar. The thick industrial poles that drop from the ceiling hold articulating arms ending in dome-shaped lights. These can be swung in multiple directions to focus on any point of the procedure table below. The table tends to be offset in the room, so that the head is closer to a wall than the foot. On the wall behind the head of the table, and sometimes hung from a ceiling-mounted boom, drop the pipes that carry the nonvolatile gases (pure oxygen, air, and nitrous oxide) used for anesthesia, in addition to a multitude of electrical outlets.

My anesthesia command center forms an arc about six feet in diameter at the head of the procedure table; I am like a pilot in a cockpit. A majority of my waking hours are spent inside this circle. I aim to control every aspect of my patient's care without having to move more than three steps in any direction. Every conceivable need for the care, safety, and comfort of my patient is contained within that arc.

Imagine the face of a clock. I am standing in the center of the face at the point the arms of the dial pivot. My patient's head rests at twelve o'clock. Along the perimeter at two o'clock, a pair of corrugated, opaque plastic tubes, each an inch in diameter, originate from the anesthesia machine on my right. When stretched, these tubes reach six feet in length. One tube flows oxygen mixed with my anesthesia gases to the patient from the machine, which stands at three o'clock, while the other removes the patient's exhaled gases. Cables for the patient monitors run along the airway circuit also from my machine to the procedure table.

The anesthesia machine is the CPU (central processing unit), the behemoth of my command center. It stands about five feet high and is about three feet square. It's loaded with metal weighing many hundreds of pounds, requiring that industrial, six-inch wheels be anchored to the sturdy iron base. A couple of drawers containing machine supplies, extra cables, and manuals sit low; the middle of the machine, the action center, contains all the dials, switches, and buttons for adjusting gas flows to switch modes for breathing. There is also a screen for monitoring the gas composition I provide and the rate and volume of the patient's breaths that I program.

On top of the machine, a shelf holds the monitor guts and the wire cables leading to the table. A flat screen displays every conceivable number and wave in a variety of sizes from font 16 to 72 and colors (red, green, and yellow). The machine fully loaded is an imposing giant.

Facing the anesthesia machine, on the left side, at waist height and rising over the gas supply tubes, extends a tube arm ending in a soft, expandable plastic bag. The bag inflates with fresh gas that, when squeezed, pushes the gas dialed in from the anesthesia machine circuit through the corrugated tubes coursing to the table. This bag allows me to breathe for my patient.

Jutting out from the machine at waist height is a

small work space that holds all the immediate needs for the case at hand. On the left are my tools for ensuring that the patient's airway remains open and unobstructed, which means he or she isn't snoring. There are plastic oral airways that look like oversized commas in a variety of sizes, endotracheal tubes (gently curved, clear plastic oversized straws that have a balloon near the end), and my laryngoscope (which lights the path through the mouth to the vocal cords). The drugs I need—the intravenous anesthesia medications, narcotics for pain relief, relaxants for temporary paralysis, and antibiotics—are arranged on the right-hand side of the table. Each syringe is labeled with color-specific tape that indicates its purpose, and each is carefully arranged with the needles pointing toward the machine.

Along the back, right edge of the table are two emergency drugs that I rarely need but always keep at the ready. The voodoo intent is to ward off evil humors. The orientation of the syringes immediately signals to me which drug I'm touching. One is a syringe of a rapidly acting paralyzing agent, succinylcholine; the other contains atropine, a drug to speed the heart. A slowing heart rate is the portent of doom. Slow heart rates progress to no heart rate and full cardiac arrest.

Behind me, at six o'clock on the imaginary clock face, is a cart that could be mistaken for a mechanic's tool chest. (It probably is one, but because it comes from a medical-supply company, the price is outrageously high.) The top of the equipment cart forms a counter that holds the not-so-urgent supplies for my procedure and provides a small work space for drawing up medications into syringes and such. Below, in six drawers of varying heights, the cart holds anything I might conceivably need for any case, from syringes and hypodermic needles to drugs and airway supplies, just as it might hold all the tools for a mechanic. The bottom drawer contains suction catheters (the most frequently forgotten item in the anesthesia setup), which are used to clear the mouth of secretions, as well as the backup to the backup, a self-inflating Ambu bag. The first self-inflating resuscitator, the Ambu bag is a plastic bag that, when squeezed, pushes a breath to a desperate patient in need. When released, it pops back to its original shape, once again filled with air. It is the last-chance piece of equipment to deliver a breath when gas flow or the electrical supply has failed.

The Ambu bag was named by a Danish anesthetist in 1957. He never revealed the source of the name. The bag became so popular and successful that the developing company renamed itself Ambu. Possible explanations of the name put forth without much proof include the acronyms for "air mask bag unit" or "artificial manual breathing unit." But in Danish—the company is based in Denmark—the word for "air" is *luft*. Not long ago, I spoke with representatives of Ambu. I pulled them aside in a meeting and asked if they knew what "Ambu" stood for. The answer? "Ambu" is an abbreviation of "ambulance."

Once, on the first day of a new resident's rotation, I demonstrated my method for anesthesia preparation, step by step. "Simply look at every drawer of the anesthesia cart before every case, and no step will be missed." The next day, this resident was scheduled to work with my mentor-turned-colleague, a person I referred to as the "Learned One," easily the brightest of my colleagues. I knew my old mentor's teaching style and the questions he asked. After all, he had taught me. I told my new resident that the Learned One would simulate a breathing-circuit disconnect—"What happens when the machine alarm sounds for the patient not receiving any breath?" he would ask—and I demonstrated what to do: when asked, immediately reach into the bottom drawer and retrieve the Ambu bag. I opened the bottom drawer and—"Damn it!"—the Ambu bag was missing. I'd screwed up.

I found a replacement, then stressed to the resident not, under any circumstance, to let this happen.

Completing the trip along the circumference of my circle, at ten o'clock a pole rising from the floor to six feet tall has hooks that hold the bags and tubing for the

IVs I use for infusions, as well as the drugs I administer during the case.

My work space looks like a little canyon, my nest.

TO PREPARE MY COMMAND CENTER, I start by connecting the airway circuit and move clockwise to the machine, ending at the cart. I quickly run my eyes from the top to the bottom drawer to ensure that every component of my anesthetic plan—Ambu bag for sure—is in one of those drawers. With one last scan, it's time to go.

There's one last item in my setup: the mother of all drugs, the resuscitator extraordinaire, the last-ditch effort to retrieve a life trying to end—epinephrine. It deserves a special place in my anesthesia nest—a place I call the "Oh Shit Shelf." It's on top of the anesthesia machine, immediately to the right of the flat-screen monitor. Whenever anyone in the room says "Oh shit," I reflexively reach high and to my right, and grab the only syringe ever placed there, the epinephrine. I am the ultimate patient advocate, and my first thought jumps to a patient in jeopardy. Epinephrine is the single strongest heart-kicking drug I have. It bumps a slowing heart rate and increases a sagging blood pressure. Along with oxygen, epinephrine is the anesthesiologist's life preserver.

One measure of an anesthesiologist's expertise might be how often the Oh Shit Shelf is accessed. Many variables beyond the anesthesiologist's skill affect the likelihood that epinephrine will be needed, including the nature and intensity of the procedure, the skill of the proceduralist, and the health of the patient. But the true measure of expertise is the ability of the anesthesiologist to manage a patient's airway, to ensure the unobstructed and rhythmic flow of breaths in and out of the patient. The first step in any critical event is to ensure that the patient is receiving adequate breaths. The paradigm is rigid: ABC—airway, breathing, circulation.

The goals are simple: rely on skill, not luck; eliminate critical incidents; and never need to use the Oh Shit Shelf.

CHAPTER 3

The Five A's

THE FIRST TWO HOURS OF MY LIFE AS A PHYSI-cian began at 8:00 a.m. on the first of July many years ago, in a lecture room with the other starting trainees in surgery. We were listening to advice—demands—on expected physician behavior, on completing medical charting in a timely manner, and on the benefit of the Office of Graduate Medical Education (that papers claiming medical malpractice would be delivered in the office and not on the floor in front of colleagues and patients). At 10:00, I was sent on to begin my practice of medicine.

By serendipity, my first rotation as a surgical intern was on the anesthesiology service. I approached the OR control desk and was greeted by the anesthesiologist charged with running the schedule for the day. At the traditional beginning of the educational year in medicine—for many physicians the first day of providing care—while the new anesthesiology residents were still attending lectures on patient care, I was released to tend to my very first patient.

"Your first case is waiting outside room 6," said the anesthesiologist in charge.

I wanted to ask him: "Are you sure you know what you're doing?"

During med school I rotated on the anesthesia service as an elective. I took it for the several hundred dollars it paid, money that I immediately dumped into stereo speakers and Bruce Springsteen's *Born to Run*.

In the hallway outside OR 6, an eighty-plus-year-old woman with a hip fractured by a fall rested on a gurney, unaware of the significance of the date. She had no idea that she would be my first patient. Fortunately for me, someone else set up the anesthesia equipment, since I had never set up a room. I didn't comprehend the concept of the command center yet, or the importance of the Oh Shit Shelf. One cursory swipe with my eyes and I decided it was time.

After transferring this unsuspecting woman to the procedure table and positioning the monitors, I pushed part of the contents of the big syringe—she was so small I needed to lessen the dose for her safety—and some of the small syringe, and turned the vaporizer to the right.

Now this woman's breathing became my responsibility. I applied the anesthesia mask to her face, squeezed the bag, and watched her chest rise and fall, the gases freely entering and leaving her lungs. Then I took a deep, relaxing breath myself.

"Go ahead," my supervising anesthesiologist instructed, referring to intubating the patient. Truth be known, during my rotation as a student I had not once intubated a patient without assistance—plenty of assistance. I grabbed the laryngoscope, similar to a flashlight, with a battery pack inside the handle, leading to a metal blade with a light at the end. I scissored her mouth open with my right thumb and index finger; slid the blade into her mouth, past her tongue, and beyond the tonsils; lifted the handle; and for the first time ever looked upon those pearly white strips on either side of the voice box, her vocal cords, and slid the plastic endotracheal tube into the trachea. I squeezed the bag on the anesthesia machine filled with oxygen and the anesthesia gases, and sighed relief when I watched my patient's chest rise with the breath I delivered. I used my stethoscope and listened to the air flow in and out of her lungs. I had complete control of her breathing.

The big syringe, the little syringe, and two clicks to the right. That's the concept of an anesthesiologist's work: the generic recipe for general anesthesia (not including one syringe to be given later—pain relief).

The procedure for general anesthesia is to push into the IV the contents of the big syringe (twenty milliliters, or a tablespoon and a third, of a rapid-acting anesthesia medication, the drug that induces the loss of consciousness); then to push the contents of the little syringe (five milliliters of a paralyzing medication); and finally, to add a gaseous anesthetic to the inhaled gases by turning the dial on the vaporizer two clicks to the right to keep the patient anesthetized. It's a little demeaning to my ego, but it's often just this easy. Behind the simplicity of the recipe is a complex mix of aims, drugs, and techniques that extends anesthesia care beyond the actual anesthesia procedure, including pre-anesthesia patient preparation and postprocedure pain relief.

The term "anesthesia," which means "without feeling," doesn't adequately encompass all of the goals of care. Since the discovery of ether, many adjunct medications have been added to the anesthesia gas to accomplish all-inclusive care. The effects that these side medications produce are what I call the "Five A's of Anesthesia":

- *Anxiolysis*, relieving stress created by an upcoming surgical procedure
- *Amnesia*, preventing memory formation during anesthesia care
- *Analgesia*, relieving pain during the procedure,

but also considered beyond the procedure
room to include postprocedure pain relief,
acute (trauma) pain relief, and chronic pain
relief
- *Akinesia*, preventing a patient's movement
during a procedure
- *Areflexia*, stopping adrenaline surge and
swings in blood pressure and heart rate while
under anesthesia

ALMOST ALL PATIENTS EXPERIENCE anxiety before
a procedure. As the twentieth century began, newly
developed barbiturates aided in comforting patients
approaching surgery. The introduction of Valium (diaz-
epam), in 1963, created a surge of anxiolysis research and
use. Over 150 million prescriptions were written for the
Valium class of antianxiety drugs (benzodiazepines) in
a single year's time.

Valium, and its short-acting younger sister Versed
(midazolam), effectively lyse the anxiety that builds as
the patient approaches the double doors. Acting within
moments of intravenous injection, the drug paints the
face of the receiver with a drowsy, intoxicated appearance. For adults, these "azepams" (as benzodiazepines
are known colloquially) have been an anesthesia game
changer for the first target of the Five A's, *anxiolysis*.

Unfortunately, children still await the arrival of a suitable drug. Most children are not given, nor will they tolerate, an IV prior to the induction of anesthesia. This constraint necessitates less invasive and more suitable routes of delivery. Starting an IV in a child is a guarantee for two minutes of crying, while the induction of anesthesia by gas takes no longer, and usually less—twenty-seven seconds by my last measure. To date, no drug has become an acceptable and effective anxiolytic for children, leaving psychotherapy (otherwise known as nonstop distraction) as the alternative. (Thank God for iPads. I remain stunned by watching a two-year-old who is not able to talk nevertheless slice watermelons on the screen with the swipe of a finger.) The syrup of Versed is used with some success, but in short cases it is associated with increased anxiety on emergence from anesthesia.

Today, there are over thirty azepams to choose from, including the sleep-inducing—and sometimes sleepwalk-inducing—Ambien and Rohypnol, street-named "roofie," otherwise known as a date rape drug.

AMNESIA, THE SECOND OF MY FIVE A's, is a natural extension of anxiolysis, since many of the stress-reducing drugs, in higher doses, induce a lack of recall. Anesthesia is lost time. There is no memory. From the moment anesthesia is induced until that point in emergence that awareness returns, a gap in time forms in the existence of the patient. If memories enter the brain during this time—debate exists as to whether the anesthesia experience prevents memory formation or memory retrieval—they never come to mind when awareness returns. The benzodiazepines help form that time hole. The drugs I use deny the formation of new memories but leave past memories intact.

Amnesia is easy to achieve, and easy to screw up. At the concentration of anesthesia gas that allows surgical stimulation, amnesia is complete. The effects of anesthesia wear off gradually, as compared to the time required for induction, and the return of memory might not be all at once. Activity, especially voices heard in the recovery room before memory restoration is complete, might be construed as having occurred in the procedure room and during a procedure. The concentration of anesthesia gas needed to prevent memory formation is much less than that required for surgery. Ensuring amnesia is easily achieved by including at least half the surgical concentration dose of the all-in-one potent anesthesia gas. Not all patients are able to tolerate anesthesia gas, and sometimes additional medications must be used. In such cases, amnesia must be a specific goal of the anesthesiologist, with medications for that purpose provided, or redosed if the procedure might outlast the actions of the amnestic drug.

Recall of events during anesthesia is possible when medications used to eliminate patient strength are not reversed in a timely fashion, as the patient is passing through a light level of anesthesia on the way to alertness but unable to react because of weakness. The patient hears everything.

Historically, anesthesia provided only calmness, and it was not uncommon for patients to be able later to recite every word spoken during a procedure. They were comfortable and relaxed and didn't complain. This was in an era prior to my career. Now, with amnesia an accepted and fundamental part of an anesthesia procedure, the routine expectation is that no patient will remember what is said and done while under anesthesia.

But I know of two patients whose memory of specific occurrences during anesthesia convinced me that they did experience recall. One procedure resulted from a displaced clamp on a large artery that caused an immediate massive blood loss. The depth of anesthesia was decreased in an effort to prevent depression of the heart's function—the gaseous anesthetic agents are known to lessen the pumping ability of the heart—and loss of blood pressure. After successful resuscitation, the patient remembered the event. Such recall is a form of locked-in syndrome, in which receptive capability remains fully intact but there is no ability to communicate.

The second instance of recall fascinates me. A four-year-old girl underwent a craniotomy for a brain tumor nearly a year before she entered into my care. The result of the tumor and surgery for its removal was twofold. First, the flow of cerebrospinal fluid (CSF) was obstructed, requiring a diversion procedure to prevent a possibly fatal condition. (CSF is the fluid cushion that prevents bumps to the head from causing the brain to strike the inner surface of the skull. CSF is constantly produced by the brain and must flow out of the skull to prevent fluid accumulation and increased pressure leading to brain injury.) Second, the surgery damaged the satiety center of her brain. This girl no longer felt full after eating a meal of any size. Possibly because of parental guilt for all that had happened to her, she wasn't discouraged from gorging, and she would eat four hamburgers at a meal or an entire box of sugary cereal in the middle of the night. There is no other way to describe her appearance: she looked like a butterball turkey, her thighs so large that her feet were pushed wide apart, and her arms so hefty that her forearms didn't touch the sides of her torso.

This girl was extremely intelligent and precocious. While she was under anesthesia for her revision surgery, and during a little OR table talk, the surgeon mentioned to me that she had experienced recall under anesthesia during the tumor resection. I asked how he knew. He said that a few days after the original craniotomy for tumor removal, during morning rounds, he had noted that this girl was on the road to recovery, fully intact, and very talkative. As he turned to leave her room, the girl had asked: "Hey Doc, what does 'get that bleeder' mean?"

"What?"

"What does 'get that bleeder' mean?"

During her operation, the surgeon, renowned for that comment—as well as "Damn it, stay with me!"—had instructed the resident assisting the surgery to use the electrocautery device to coagulate an open blood vessel. There was no possible way that a four-year-old would know this medical lingo.

I made sure that she would experience no recall during my care.

AN ABSOLUTELY MOTIONLESS PATIENT is a must during critical moments of a procedure. For a cardiologist about to burn an electrical path gone awry in a heart, or a radiologist about to dilate a narrowed blood vessel in the abdomen, or a neurosurgeon about to place a clip on a brain aneurysm that is ready to burst at any second, a movement as small as a millimeter could adversely affect the patient's life.

Ensuring total patient stillness during these essential

points in procedures has long been a goal. Its eventual means was first established centuries ago, when European explorers observed South American Indians while hunting. Using blowguns and arrows tipped in wourali poison, natives were able to bring down game animals. The method was called the Flying Death. Eventually, purification of the poison resulted in the discovery of curare, a reversible paralyzing drug first used in anesthesia for *akinesia* (muscle relaxation) in the 1940s. With a patient pharmacologically paralyzed, unable to voluntarily move any muscle in the body, the depth of gas anesthesia could be decreased, thereby allowing less depression of the heart and shorter time for emergence. Operating on a motionless target is critical in some procedures, such as brain surgery, but it's also a benefit in others, leading to the oft-heard surgeon's plea: "I need more relaxation."

THE SEARCH FOR *ANALGESIA*, the relief of pain, dates back many thousands of years. Today's potent anesthesia gases, the one-stop anesthetic, completely obliterate the presence of pain during a procedure. At some point the procedure is complete and my all-in-one gas must be turned off for the patient to return to consciousness. Another method of continuing analgesia becomes necessary. The crude methods of chewing tree bark and leaves and drinking the juice of flower seedpods evolved into the sophisticated pain relief medications that are purified and discovered in pharmaceutical and university laboratories. Willow bark evolved into aspirin, coca leaves into cocaine, and the opium-containing poppy into morphine.

In the Andes, the Incas discovered that the coca leaf holds many beneficial effects, including pleasure and numbness. Trephination—that is, drilling holes in the head, one of the earliest-known surgical procedures—was used to treat seizures, headaches, and mental ills. Inca shamans chewed coca leaves and spit into the head wounds. The cocaine contained within the leaves numbed the area and, as an added benefit, constricted blood vessels, lessening blood loss.

William Halsted, a Johns Hopkins surgeon of the 1880s and arguably the most influential modern-era surgeon, applied the concept of antisepsis to surgery, thereby vastly increasing the scope of invasive procedures. His meticulous technique required anesthesia and provided safe operative conditions. Frances Burney described at length her harrowing experience during a mastectomy without anesthesia. Few would accept Halsted's procedures without anesthesia.

Centuries after the Inca shaman, as surgery became the norm for immediate cure, cocaine was purified and found to block nerve impulses from reaching the spinal cord and brain. Halsted took a hypodermic needle to the nerve innervating the upper jaw, injecting cocaine and creating numbness that allowed for painless oral procedures. He then found personal pleasure in cocaine and used it for recreational purpose, as did the eminent psychiatrist Sigmund Freud. Toward the end of the nineteenth century, cocaine was applied topically to the eye. Finally, cocaine was applied directly to the spinal cord to provide prolonged loss of sensation to the lower body without mental sedation— spinal anesthesia and epidural anesthesia, depending on how deep the needle is inserted between and beneath the vertebrae. Regional anesthesia was born.

Pain starts by activating a receptor at the source of the injury that transmits information to the brain as an electrical signal conducted by nerves. An insulator, the fatty membrane myelin, acts like the plastic coating on an electrical wire, preventing the loss of the signal to adjacent tissue as it travels. Breaks in the myelin called "nodes of Ranvier" enhance the speed of transmission of the pain signal by allowing the signal to jump quickly from node to node along the nerve. This rapid transmission is known as "saltatory conduction." Cocaine enters these nodes, blocking the signal jumps and thus the transmission of the pain signal beyond that point.

The 1920s witnessed not only Prohibition, banning alcohol, but also legislation that formally banned narcotics. No longer could Mrs. Winslow's Soothing Syrup—a morphine-laced drink offered as relief for teething toddlers—be purchased on a whim. Sales now required a prescription, creating a great example of the law of unintended consequences by leading to the formation of the black market in drugs. Narcotics remain a mainstay of my analgesia efforts, especially in treating acute pain caused by surgery or trauma. Analgesia remains sacrosanct to the anesthesiologist. But a fine line divides alleviating pain from abusing drugs, as demonstrated by the forty-eight thousand deaths that result from substance-use disorder every year in this country.

In the decades that followed the discovery of the medicinal use of cocaine, chemists, pharmacologists, and physicians dissected the coca leaf to its most basic active compounds. The knowledge gained led to the introduction of a multitude of local-anesthesia drugs. Perhaps the most widely known is lidocaine, which, in addition to numbing nerves, stabilizes the heart from erratic beats. Research on new medications has slowed to a crawl, however; the last significant drug was added twenty years ago. Despite the lack of new drugs, the use of regional anesthesia is expanding, with better and easier imaging techniques making it possible to accurately place current drugs on more nerves.

AREFLEXIA IS A CONCEPT that I still struggle with, because it is like lassoing a cloud. It consists of adjusting the depth of anesthesia, controlling the patient's blood volume, and adding drugs that alter heart rate and blood pressure, while taking into consideration the procedure and the patient's health.

Controlling the heart rate and blood pressure of an anesthetized patient requires assessing many variables, and the relationship between the two is complex. For a sick heart, one with narrowed coronary arteries, slowing the heart rate is beneficial because it lessens the work of the heart and improves blood flow to the heart muscle. The opposite is true with a burst brain aneurysm, in which case maintaining a hyperdynamic state—keeping the blood pressure high—is thought to improve brain blood flow and survival. With a young patient, a faster heart rate is the goal. Some surgeries, such as correcting curvature of the spine, are prone to extensive blood loss, which can be limited if the blood pressure is intentionally lowered independent of the heart rate.

DECONSTRUCTING ANESTHESIA FROM THE all-in-one ether-filled orb into the Five A's expands the responsibilities of the anesthesiologist to include the preprocedure relief of anxiety and the postprocedure continued relief of pain. Then, targeting one of the individual A's provides optimal conditions and improved outcome for more complex and delicate procedures, such as inducing akinesia to ensure an absolutely still patient as the clip is applied on a cerebral artery aneurysm. Over the course of my career, perhaps the greatest change and also my greatest stressor has been the increase in the number of critically ill patients undergoing procedures and requiring anesthesia care, with many of these procedures life threatening and many of these patients at risk for death. That cloud must be lassoed so that areflexia can be understood and precisely controlled for optimal patient outcome.

CHAPTER 4

Railroad Tracks

I N MY ANESTHESIOLOGIST DREAMS I SEE RAILROAD tracks. Two rails connected by regularly repeating ties, running perfectly straight, parallel, and unobstructed, and drifting across my view from left to right until reaching the horizon. This is the landscape of my ideal anesthesia record. The rails are the ticks and dots representing the patient's blood pressure and heart rate as recorded during the progress of my anesthesia care, without variance. The ties represent time, with five-minute intervals the convention for recording vital signs. In the best-case scenario, these marks form parallel lines with no slope, marching horizontally across the anesthesia record. The ultimate, stable patient.

From "Your patient is ready"—the comment made at that first painless surgery in 1846 and still used today—to "I'm finished," my goal is a period of stability, of unvarying vital signs, of something like boredom. The well-being and survival of my patients depend on stable vital signs throughout the procedure and after. A spike in blood pressure might burst a fragile blood vessel; a rise in heart rate could push a failing heart past its limitations. My objective is to finish with an anesthesia record that marks the vital signs as flat and straight across the chart—my "railroad tracks"—a plan well accomplished. This is easier to write about than to put into practice.

From the earliest anesthesia records to the present, the sheet used for logging all the relevant data has featured a grid of boxes filling the central portion of the page, providing space to record the ticks and dots of the vital signs. The space surrounding the grid lists demographic data, including the patient identifier, weight, procedure, reason for the procedure, allergies, and a description of the anesthesia technique used for the procedure, drugs included.

Fifty years after the first ever recorded surgery performed with the patient under the effects of inhaled ether, the anesthesia record was developed. In 1905, a group of physicians with a common interest in the science and art of anesthesia held their first meeting, which led to the formation of the Long Island Society of Anesthetists, an organization that expanded throughout the state of New York and then the country, eventually changing its name in 1936 to the American Society of Anesthesiologists. Before that, medical students and surgeons were the ones who administered the ether or, in some cases, especially in Europe, chloroform.

John Snow administered chloroform to Queen Victoria for painless childbirth in 1853. In the 1890s, a pair of medical students, one about to become one of the world's early great neurosurgeons (Harvey Cushing) and the other a leader in the organization of outcome studies in medicine (E. A. Codman) astutely noticed, while providing ether to patients, that certain bad outcomes followed trends in vital signs that veered from the straight and parallel nature of the railroad tracks. Cushing administered ether to a patient who vomited and aspirated the stomach's contents into the lungs shortly after losing consciousness, and then promptly died. "There was a sudden great gush of fluid from the patient's mouth," he wrote, "most of which he inhaled and he died."

Cushing and Codman soon developed a system to record the vital signs that were measurable at the time. From the very first dot and the very first tick placed on the first anesthesia card, Cushing's and Codman's intent was not to record history, but rather to predict and prevent poor outcomes.

Though perhaps a bit more formalized and organized now, the grid remains split into sections with many more vital signs that, as they became measurable, gained a place in the record. The squares of the lower section provide the space to document the patient's vital signs in relation to elapsed time. A heart rate of one hundred beats per minute, measured fifteen minutes after the start of anesthesia, would be logged as a dot at the tenth box up and three boxes to the right of the start line. But one set of vital signs is not a trend. Within any ten-minute segment, three sets of vital signs, each with five measurements—blood pressure, heart rate, respiratory rate, oxygen saturation, and end-tidal carbon dioxide (exhaled gas)—provide enough information to make clinical decisions and adjustments.

The vital signs available to measure at the time Cushing and Codman introduced anesthesia cards were limited to heart and breathing rates. Both were measures of quantity; they didn't measure the quality of the heartbeat or the amount of air exchanged with each breath. The average body contains over sixty thousand miles of conduit, blood vessels, that carry nutrients to and waste from all the cells of our bodies. The heart beats on average 115,000 times per day, transporting ninety-two hundred gallons of blood. In a day's time, the average person takes over twenty-three thousand breaths. But those numbers fail to tell the story or ensure health. Today, the goal of every anesthesiologist is to make sure that the heart functions well enough to propel blood throughout all the blood vessels and to carry the fuel (oxygen) that is loaded in the lungs and intended to keep the powerhouses (mitochondria) of the furthest cell from any blood vessel energized and functioning properly.

In 1901, Willem Einthoven used generated-for-consumer electricity to measure generated-by-the-heart electricity, producing the first electrocardiogram (ECG, sometimes called an EKG because it is derived from the German word *Elektrokardiogramm*). Two decades later, the innovation would earn him a Nobel Prize in Physiology or Medicine. In 1920, anesthesiology recognized that the flow of electricity through the heart not only provides information on rate and rhythm, but can discern a failing heart. Unfortunately, the Einthoven ECG machine, at six hundred pounds, could not easily be used for all anesthetized patients.

Monitoring the heart by electrocardiogram grew slowly in popularity, perhaps because of a lack of certain technological advances, paired with the dangerous presence of flammable anesthesia gases in most surgical situations. The routine use of ECG monitoring during surgery was finally adopted in 1960. Although the call for standardized anesthesia records began in 1923, not until 1985 did standards for monitoring, including use of the ECG, become accepted by the American Society of Anesthesiologists. Eventually, monitors for inhaled and exhaled gases were added.

RECORDING A RAILROAD TRACK anesthesia record is easy in concept but challenging in practice. Finding a status quo in vital signs requires anticipating fluctuating body stresses during the progress of a procedure and adjusting the depth of anesthesia before allowing the tracks to take on the appearance of a roller coaster streaking up and down. The stress of making incisions in the skin, Bovie-cauterizing breached blood vessels (William Bovie was a physicist and colleague of Cushing, who in 1926 developed an electric wand used to coagulate blood vessels and decrease blood loss), and cutting bone causes catecholamine surges, which result in increased heart rates and spiking blood pressure. These vital-sign changes, in turn, necessitate more profound anesthesia. As time passes with little stimulation, such as while waiting for an X-ray or pathology report—the downtime during a procedure that is known as a DUA (discussion under anesthesia)—the patient requires a lessened depth of anesthesia.

Whereas some surgeons are models of efficiency, others thrive on DUAs. Hand surgery is performed with the patient's arm extended out to the side and the hand placed on a table attached to the OR table. The surgeon sits during the procedure, as at the dinner table. In the middle of a case, one particular hand surgeon I knew—despite his pokiness, a favorite of mine—would put his surgical instruments down, rest his elbows on the hand table, clasp his hands at mouth level, and dive into a full-blown DUA regaling all in the room with stories from a long and eventful career. I learned to remain very quiet while performing anesthesia on his patients, for if I spoke and prompted him to launch into a DUA, the OR staff would glare at me.

"Calm." "Lull." "Doldrums." These words are my friends; they describe my goal for the "interlude," the maintenance period of an anesthetic, the time between the induction of anesthesia and emergence. No swaying patient stresses, causing swaying vital signs as the procedure progresses. Steady anesthesia depth. My records appearing as railroad tracks.

It is the interlude that gives an anesthesiologist the opportunity to grow from competent to great. During the induction and emergence periods, all actions on the patient originate with the anesthesiologist. During the interlude, the actions on the patient turn from the anesthesiologist to the proceduralist. An anesthesiologist who sits behind the surgical drapes can only react to the changes in a patient's vital signs that result from the procedure. But an anesthesiologist who stands can observe all that happens and learn to become proactive, monitoring the moves and competence of the proceduralist. That's the period when I watch with care the skill of the proceduralist. If I learn to predict the surgeon's next action and the effect on the patient, I can move to prevent unfavorable changes in my patient before they become problematic.

This is the time when I observe the surgeon's qualities and abilities. It was during an interlude that I recognized perhaps the greatest surgeon I've ever worked with. Watching Casey perform surgery was like seeing an old master's brushstrokes. Casey's hands flowed through a surgical field like an artist's brush across a canvas. I saw him think as he developed new uses for the standard instruments. As Casey's hands with knife and scissors dissected tissues, he laid out the anatomy for any observer to witness. He was a surgical magician. He taught me anatomy and technique.

VITAL SIGNS FOR SOME CASES are prone to roller-coaster rises and plunges. One of my most memorable cases, perhaps my most difficult case, was providing anesthesia for a man's liver tumor resection. A tricked-out IV pole—a rapid intravenous infusion device— helped save the patient. The liver is notoriously vascular, and tumors only worsen the situation. As I watched, I noticed that every time the surgeon's hands touched the

liver, blood welled out of the belly. Besting the surgeon's attempt to drain the patient's body of blood, the
rapid-infusing IV device replaced the patient's blood
volume and sustained his blood pressure throughout
this marathon procedure.

The infuser consists of a three-by-five-inch box
bolted to the IV pole at a height of about five feet. The
main unit contains a liquid-based fluid warmer and
pump that pressurizes two separate chambers hung
from above. A bellows in these chambers inflates and
squeezes fluid out of the plastic bag under pressure,
sending the warmed fluids through the IV tubing to the
patient at a rate that exceeds most surgeons' ability at
bloodletting.

Shortly before the advent of anesthesia, the practice
of providing fluid directly to the veins was begun to
offset the dehydration resulting from cholera's profuse
watery diarrhea, a frequent cause of death. Sterile fluids
for intravenous administration were introduced in the
1930s, and later came improved catheters and tubing.
The rapid infuser allows fluids and blood to flow to the
patient at high rates while both warming the fluid and
preventing the inadvertent injection of air, which can
create a lethal airlock inside the heart.

During the unforgettable tumor resection I watched,
I lost count of how many times I saw the surgeon's hands
touch the man's liver, or of how many units of blood I

transfused. Despite my best efforts, the patient's vital signs drifted low. I added medications that boosted the heart and squeezed the blood vessels, but still the trend was in the wrong direction. After a few hours of this battle, and with the patient's temperature dropping, I turned to the surgeon and told him: "Put your instruments down and step back from the table. Go away; I'll call you back when I'm ready."

I needed time to adequately resuscitate the patient, to replenish the patient's blood volume, and to correct the labs. The surgeon listened to me. A half hour later, I called him back into the OR—the patient now stable, all the numbers normalized (the temperature was still a little low, but I would correct that more slowly)—and the anesthesia record returned to railroad tracks. The patient survived the procedure, but I don't know whether he survived the tumor over the long term. The case ended late at night, and I couldn't get to bed fast enough. This was a rocky interlude saved by technology. In that OR, I was the one who was drained—of energy.

The interlude is normally uneventful, but always uncomfortable and, in some cases, nerve-racking. But in my anesthesia nest, I can always touch my patient, which provides me some sense of security.

CHAPTER 5

Fear of the Mask

FIRST MET AMY BY THE SOLES OF HER FEET.

She'd been assigned bed space 4 in the "sandbox," a space adjacent to the operating rooms that serves as a pre-anesthesia staging area in the pediatric center. A nursing station overlooks the area, which is about twenty feet square—big enough for three patient carts. The front is open, while the back is a wall of windows, allowing freely flowing light and a view. Three equal-sized spaces occupy the sides, with carts separated by ceiling-hung sliding drapes that stop about a foot from the floor.

Privacy is nil. The edges of the drapes never close completely, and there is no way to have a private discussion. But the sense of all being in this together pervades. No one ever complains about the lack of privacy, and almost all are cheery, proving that bigger isn't always better. This area is the result of the old building having been remodeled multiple times, in an attempt to maintain pace with evolving technology and the changing needs of health care.

Amy was ten, and I had been warned that she was a little "high-strung." Armed with this information, I raced into the sandbox with the tails of my gray lab coat flying behind me, my right hand as usual wrapped around a rolled-up paper surgical hat just removed and stuffed into the waist pocket.

Entering Amy's space required that I walk past it to the windows, to find the edge of the privacy curtain. I pulled it back and found her cart empty. Aside from some personal belongings, there was no Amy, and no family. In no more than a second, possibly two, before I could even scan the area, I heard her mother.

"Amy! Amy, you get back in here! Amy!" There was a momentary pause as her mother caught her breath. "Amy, you're embarrassing me! Amy!"

I cocked my head to the right but still saw nobody. Then I leaned to the left and saw Amy's mother squatting on the other side of the cart.

"Amy, you get back in here!" her mother said.

The bottom of Amy's feet came into view, her toes to the floor. In an infantryman's low crawl position, with her wrists hidden under her chest, her elbows jutting out to her sides alternating thrusts while staying low to the ground, she was attempting to snake her body along the floor, as if maneuvering under barbed wire. My gaze continued up Amy's back, which ended abruptly at her shoulders, as her head and neck disappeared under the curtain. She was trying to slide into the next patient space. Her mother's hands were firmly wrapped around Amy's ankles, preventing the escape, and I heard her mother grunt as, with a great heave, Amy was yanked back to her designated space. Her mother then hoisted Amy up and plopped her on the cart.

The pleading look that Amy shot at her mother was unmistakable. Amy was prepared to do anything to get out of her situation.

"Hi. I'm Doctor Jay. I'm the anesthesiologist."

"Oh boy," was Amy's mother's only response.

ANESTHESIOLOGY IS A SPECIALTY that is normally nontherapeutic. I provide the assistance that allows the treatment. The "Do no harm" motto of Hippocrates weighs heavily on me. Providing safe care for the child is the only issue more important than easing anxiety before anesthesia is induced. If a perfect pre-anesthesia sedative medication existed, every child—and parents on an as-needed basis (and they almost always need it)—would receive it. There is no such medication. And, children often are unable to swallow pills. The alternative is an elixir that is frequently spit out or vomited at the start of anesthesia. For some kids, the only alternative is an injection of a sedative, and then they cry for the two minutes it would take me to induce anesthesia by mask and gas. I prefer to use little premedication and lots of nonstop talking until the child is unconscious.

My patients are not always willing participants, and I'm often viewed as the one forcing their participation. I'm the last physician a patient interacts with prior to losing consciousness. Placing a mask for the induction of anesthesia, sometimes against the will of the child, can be tricky. Older children and adults reflexively push the mask away when it is first introduced. Claustrophobia and a feeling of suffocation, despite plenty of gas flow, are the common complaints. The anesthesia mask may induce fright; a phobia of the mask, an unreasonable fear, may develop.

Phobias hold lifelong implications, of course. And it's reasonable to assume that the face mask I use to induce anesthesia could create a phobia, as it is the last memory prior to any invasive and extensive surgery. But I have never received a complaint that the anesthesia experience negatively and irrevocably altered a child's behavior. And no psychiatrist, to the best of my knowledge, has ever come forward questioning the psychological impact of my anesthesia care on a child patient.

My concern motivated me to study children with anesthesia mask phobias. Rarely does a child volunteer fear of the anesthesia face mask before a procedure. And not one of the phobic children I've encountered has ever been able to describe the moment—or the procedure— when the fear developed. Mention receiving an injection, a shot, before anesthesia, and most children freak. Fear of the hypodermic needle is real and pervasive. When given the option, children always choose "going to sleep with the mask" over an injection. Anesthesiologist colleagues (not parents) have told me about alterations in certain children's behavior following a procedure that required anesthesia: refusing to accept any lollipop after being given a lollipop sedative; rejecting all cherry popsicles after use of a cherry-scented mask; insisting on having a nightlight on at all times in the bedroom. In the last case, I remain curious about whether the boy in question was told "Lights out" at the induction of anesthesia.

SINCE EVERY STEP TAKEN toward a procedure room increases the anxiety of my patient, my goal is to shorten or disguise the time from that first step until my anesthesia coma is induced. Distraction is a major tool, and maintaining an ability to distract keeps me young, or at least requires that I stay current and informed. Bands, books, TV shows, personalities, the latest news and, even better, the latest gossip—I have to know it all.

I met Adam when he was twelve, that miserable middle-school age. Tweeners are never easy to connect with. Add a necessary medical procedure on a testicle or penis, and the situation turns unbearable for all. The time from first meeting the tweener until I steal consciousness cannot be too short.

At that age, life fluctuates with every passing second. A changing body collides with expanding maturity and raging hormones, and the result does not always make sense. Whispers from friends and classmates as a procedure approaches outpace these growing minds, and this is the age of proliferating horror stories. "Did you hear about the boy who had a mole removed? He was an honors student before, but not anymore." Such utter nonsense is passed between these kids, only adding to their anxiety.

For Adam, making matters worse (or at least more sensitive), the procedure would be on one of the most intimate parts of his body. The left side of his scrotum was twice the size of his right side. As a boy develops in the womb, his testicles form in the belly, then descend into the scrotum. This path does not always seal tightly or strongly, and sometimes the lining of the belly—the peritoneum—slides through this tract, causing a bulge often seen in the scrotum; the result is an inguinal hernia. A loop of bowel might become trapped, or perito-neal fluid might accumulate—the latter known as a hydrocele. Surgery is required to close the path.

Sometimes the testicle doesn't reach its final destination in the scrotum (an undescended testicle), and surgery (an orchidopexy) is needed to make a path and pass it to where it belongs. For twelve-year-old Adam, this surgery on his manhood was not a pleasant thought—and embarrassing, to say the least.

Adam was a bit chunky—a challenge for placing a pain relief block accurately. He didn't appear athletic, and he spoke technogeek. As soon as we were out of hearing range from his mother, seeking to distract him on the way to the OR, I wanted to ask my usual question concerning girlfriends but was leery; I was unsure if he had reached this stage. I asked anyway.

"So Adam, what's the name of your girlfriend?" I was more direct than usual.

"I don't have a girlfriend."

"What grade are you in?"

"Sixth."

"Sure, you have a girlfriend. What's her name?"

"I don't have a girlfriend."

We approached the OR door without Adam having a clue how far we had traveled.

"Adam, I think you do have a girlfriend. What's her name?"

"I don't have a girlfriend. But if I did, I'd call her Madame X."

"Aha! Busted." Now inside the OR, I stopped his cart short of the procedure table.

"Adam, first, I know you have a girlfriend. And second, you need to know that what's said in the OR stays in the OR. Nobody but us will ever know. Just fess up."

The OR nurses chimed in: "Yes, Adam. That's right. Nothing leaves the OR. And he's not going to leave you alone."

"Adam, just give it up. You'll feel so much better. Just tell us her name."

There was a pause. Then he said: "Sarah."

Adam, for all I knew, might have made this name up just to end the discussion. Still, with my goal accomplished, Adam was distracted and quickly asleep, and calmly so.

Surgery was uneventful. I placed a nerve block for postanesthesia pain relief, and Adam left the OR with a balanced scrotum of equal size.

An hour later, I walked over to the outpatient center to check on him and determine the effectiveness of my nerve block. I saw him sitting up on his cart watching TV. I knew my block had worked. As I entered his room, where his mother sat near his feet, Adam noticed me, and his eyes grew wide. A worried look came over his face.

"Remember," his voice cracked.

"Remember what?"

"Remember," he said again, slightly louder and more demanding.

"Remember *what*?" I said a bit more firmly, and gave him a subtle wink.

"Remem . . . Ohhhh."

"What are you two talking about?" his mother asked.

What's said in the OR stays in the OR. With that I began to turn away, but caught a glimpse of the wink that Adam sent back my way.

If Adam looked back on his trip to the OR, he would be unable to estimate how long in distance or time his journey to anesthesia was. Distraction had removed him from his surroundings and anxiety. Adam didn't need any medication before heading to his procedure—especially a medication that could potentially make recovery more difficult.

HIGH-STRUNG AMY SUFFERED from fibular hemimelia, a bony defect of her leg. The lower leg consists of two bones, the tibia (shinbone) and the fibula, a narrow bone to the lateral side. The fibula, fragile in appearance and non-weight-bearing, is the lateral buttress of the ankle, stabilizing the ankle and allowing the foot to plant flatly when walking. Hemimelia is the congenital absence of one of the bones of a distal limb—in Amy's case, the fibula. Without it, lateral support for the ankle is lacking and the foot rolls in, the sole facing sideways, the inner ankle contacting the floor. Amy had what is now known as longitudinal hypoplasia of the lower extremity. Simply stated, for as yet unknown reasons, her fibula had not fully developed. She was left with a bowed shinbone and an unstable foot. I didn't notice any crutches near Amy, so I'm not sure how Amy got around.

Injustices of life come in many manners and on many levels. Twenty thousand single gene defects, give or take a few, cause a wide array of abnormalities, and then there are all the congenital defects, which are not genetically inherited. Amy's defect was nasty and disfiguring—but not necessarily damning. Amy's life would be altered, but not shortened. She would not be a ballerina, but she might write the score. She might not be a runner, but she could be a sports physician. Neither her life nor her intellect was ever at risk. She required multiple surgeries because the plan was to attempt to salvage her leg, as opposed to amputating it.

Prior to our meeting, Amy had undergone multiple surgeries and was well aware of her circumstances. I never drew out whether she had had a negative experience. She knew she was going to have another surgery that would cause pain and discomfort, and once again would be incapacitated for a prolonged time. She was spunky, insightful, and right. This was happening again.

"Oh my God; I can't believe I'm doing this." That was

Amy's response to my every question, and always uttered in a high, squeaky voice. That was her dirge even when I didn't ask a question.

Yet Amy surprised me by how cooperative she became. She was indeed "a little high-strung," but she answered all my questions and allowed my examination without complaint.

"I'm sorry. I'm so embarrassed," her mother lamented.

I looked across the cart, over Amy, to her mother, arched my eyebrows high, and flashed a wide, closed-mouth smile. "We'll be OK," I said. I couldn't hold Amy responsible for all she had been through, but her mother seemed to be holding herself too responsible for Amy's defect and behavior.

Then Amy, her mother, and I discussed the possible sedation medications I could prescribe before her surgery, to help her become calm. I explained that the best way to ensure the effectiveness of the medication was through an intramuscular injection.

"A shot!"

Amy quickly pointed out that the operating room was a better alternative to receiving a shot.

"No way. I'm not getting a shot." There was no doubt in my mind that if I approached her with a hypodermic needle in hand, I'd be treated to the vision of Amy's back becoming smaller and smaller as she ran away as best as she could.

"OK. No shot. The other choice is to drink some medicine. We dose it safely, so you still might remember the start of the anesthesia. There are a couple of side effects, just to inform you fully."

"What are they?" asked Amy's mother.

"First, there's a chance the anxiety might be worse when she emerges from anesthesia."

"She can become *more* anxious?" Her mother spoke in a tone somewhere near disbelief.

"Second, there's a higher incidence of nausea and vomiting after the procedure."

"You mean I might throw up?" Amy said.

Amy's mother was quick to point out that Amy had not vomited before. Together they chose to forgo any pre-anesthesia medication.

While I could reassure her mother that all would be well, there was no assuring Amy of anything of this nature.

"Oh my God; I can't believe I'm doing this. Oh my God . . . "

"OK. Time to go," I said.

Amy, sitting on the cart with its back elevated so that she could relax and see more than the ceiling, turned to her left and grabbed her mother's arm. "Mommy, please. Mommy!"

"Amy, the sooner this happens, the sooner it will be over. You'll be fine."

Amy's mother pried her arm free, and without further hesitation, I rolled the cart out of the sandbox and toward the operating room. Amy had seemingly resigned herself to her fate. There were no more attempts to delay the procedure; no jumping-off-the-cart escape attempts. She sat back, crossed her arms across her chest, and pouted.

But every ten feet or so as the cart rolled along the hall to the OR, Amy leaned forward and pleaded: "Oh my God; I can't believe I'm doing this. Oh my God; I can't believe I'm doing this." Unfortunately—for Amy more than me—her case was scheduled in the OR that was farthest from the sandbox, prolonging the time until she was anesthetized. So I heard her moan over and over again. It was always a couplet. Once in the room, I moved like lightning and Amy was quickly anesthetized. To my surprise, Amy accepted the anesthesia mask. There were just more moans.

The surgical goal was to start lengthening the tibia, and this wasn't the final corrective procedure. Amy would return to the OR.

After Amy was awake in the recovery unit, I spoke with her mother. I wasn't displeased with myself, but I wasn't happy either. I felt I hadn't done enough to help Amy with her anxiety. You might have knocked me over with a feather when her mother asked if I would be amenable to caring for Amy when she returned for another procedure.

A strange thought coursed through my mind: If this was, in her mother's mind, a success, what might Amy have been like before?

"We need to do something better to make this easier for Amy next time," I said.

"Thank you. Thank you. Thank you."

Amy's anxiety bothered me. My goal was to alleviate the stress, have a peaceful separation from her mother, create a state of calm on the ride back to the OR, and—maybe asking a bit too much—have Amy smile. To me, it's disconcerting, bordering on painful, when a patient is noticeably stressed during this time. I was driven to make Amy's next experience better.

Several months later, I received a phone call from Amy's mother. Her recovery from our first procedure together had been fine; there were no complications, and the pain was not bad. This time, Amy and her family allowed me to order some medication that could be used before they entered the hospital. Amy swallowed a Valium tab as they drove into the hospital parking garage; the timing was intended to allow the medication to take effect and calm Amy at the point she most needed it, as the OR time approached. There were no under-the-curtain crawls this time. Still, there was a litany of "Oh my God; I can't believe I'm doing this. Oh my God; I can't believe I'm doing this."

"Amy, how's school?"

"It's fine. Oh my God; I can't believe I'm doing this."

"Did you have a nice holiday?"

"Yeah. Oh my God; I can't believe I'm doing this."

"Do you have movies picked out to watch after the surgery?"

"No. Oh my God; I can't believe I'm doing this. Oh my God; I can't believe I'm doing this."

Once again, we slipped into the OR and anesthesia was quickly induced.

Amy emerged nicely from anesthesia, as before. I spoke with her mom, and once again she asked if I would care for Amy for her next procedure. I came to really like Amy, and I enjoyed caring for her. When she was to have another surgery, I rehearsed in my mind how our interaction would go. Would I finally break the chant?

The following year, now a newbie teen, Amy received her Valium tab thirty minutes out from arriving at the hospital.

Once again, on the way back to the operating room: "Oh my God; I can't believe I'm doing this. Oh my God; I can't believe I'm doing this."

"Amy, has anything bad ever happened to you under my care?"

"No. Oh my God; I can't believe I'm doing this."

"Then why do you keep saying, 'Oh my God'?"

"I don't know. Oh my God; I can't believe I'm doing this. Oh my God; I can't believe I'm doing this."

Before her next procedure, Amy, now fourteen, took two tabs of Valium. Once again the litany began: "Oh my God; I can't believe I'm doing this. Oh my God; I can't believe I'm doing this."

"Amy, does chanting 'Oh my God, I can't believe I'm doing this,' make you feel better?"

"I don't know. Maybe. I guess so. Oh my God; I can't believe I'm doing this. Oh my God; I can't believe I'm doing this."

The final time I cared for Amy, by then sixteen and a junior in high school, the double dose of Valium was taken before leaving home. Now maturing into a woman, her voice had mellowed and was no longer squeaky. Between all of her chants on our trips to the OR, I found Amy to be a really sweet girl.

We pulled out of the sandbox after she kissed her mom, and once we had passed through the double doors into the hallway to the OR, I asked: "Amy, how's school going? I bet you're a good student."

"Yeah, I am. It's going fine. Oh my God; I can't believe I'm doing this. Oh my God; I can't believe I'm doing this."

I stopped the cart in the middle of the hall, moved from the head of the cart to her side, leaned over the railing, and asked: "Amy, do you have a boyfriend?"

"Yes, I do."

"What's his name?"

"John."

"Amy, has John ever seen this side of you?"

She paused for a moment, then began chuckling. "No. Thank God." She laughed until she was anesthetized. Finally there was no more "Oh my God; I can't believe I'm doing this." In the end, long after I had hoped to, I achieved my goal; I got the smile.

Nearly running with Amy on her cart and down the hall wasn't fast enough to lessen her anxiety. Pharmacology had failed both Amy and me. Finally, the art of distraction came through. Only after I managed to look at her behavior through another's eyes, her boyfriend's, did Amy objectively see herself.

I was proud of having helped her overcome her anxiety. But I never got the chance to watch Amy walk. This is a shortcoming of anesthesiology. Even though I provide care during the most stressful, critical, and intense time, I don't get to see patients when they return for clinic visits after healing.

DISTRACTION IS NOT WITHOUT its problems. I regret the day I conned one certain teen into revealing the name of her boyfriend.

Poland syndrome is another of those fortunately rare but obnoxious defects that doesn't alter the length of life, but does alter appearance. Named after the London physician who first described the defect, it consists of one-sided absence of a chest muscle, the pectoralis. In girls, the overlying breast doesn't develop. This is the one no-doubt-about-it indication for a teenager to receive an implant to balance breast size.

Nikki defined the image of a high school cheerleader. She was pretty and thin, and in her hospital gown there was no hint of any breast malformation. It wasn't her looks, though, that made her stand out. It was her smiling personality. She was one of those girls it was impossible not to like.

On the trip to the OR, I asked Nikki about a boyfriend, but she refused to surrender his name.

Then I veered onto a path I regret, to this day. It's a question of integrity, and I compromised my own when I mentioned that the drug she was about to receive is known as truth serum.

Nikki said nothing more as we entered the OR. When I was just about to inject the anesthesia induction medication, and with no more talk, Nikki sat up on the table and announced to all not just the name of her boyfriend, but also his phone number. For all I know, the name she gave might have been contrived and the phone number might have reached a you've-been-dumped recording. As Nikki emerged from anesthesia, she said nothing—just smiled.

But I'll always be haunted by the fact that my attempt

at pre-anesthesia distraction seemed only to make Nikki more uncomfortable, not more calm.

FEW PROCEDURES INCITE MORE fear and anxiety than those on the penis.

Approaching double-digit age, a boy named Sam required a penile rehab to remove a hood retained from a botched neonatal circumcision. It seemed odd that the family had waited so long before deciding on the procedure. Sam's age should have prompted me to be more suspicious. Why wait this long? Only the inhibitions of his parents could be the answer.

I entered his room to prepare him for anesthesia and noted his mother sitting to his right, near his head. Before I could say a single word, before I was even near the foot of her son's bed, she burst into tears that streamed from her eyes and dripped off her cheeks. Her husband, standing against the wall on the opposite side of the boy, dropped his head in an oh-boy-here-we-go manner. She looked straight at me and declared that she was going to accompany her son to the operating room for the induction of anesthesia. "The surgeon said it was OK," she said.

"No." Just a single, calmly and firmly spoken word.

"What?"

"No," I repeated.

"Why not?"

"Because it's obvious."

"It's obvious?"

"Very much so."

"Really?"

"Yes."

I didn't need the grief from a parent's unchecked emotions following me into the OR. And she didn't need the extra stress and uncontrolled emotions.

She paused momentarily, considering my words. She didn't question *what* was obvious. She turned slightly to Sam, then raised her left arm over his chest and pointed to her husband, while returning her gaze to me.

"Then he'll go." She understood her inability to control herself. Her emotions preceded her every word.

I recognized my limits; I wasn't going to stretch my luck with this mother any further. She had lost the initial and most significant part of her negotiation. She wasn't going back to the OR with her son, but she also wasn't about to lose any more. The funny thing is that her son showed no anxiety, quietly focused on the video game he was playing throughout all of this.

A few minutes later, Sam's father, who had not yet said a single word, put on our bunny suit (a white paper neck-to-toe gown, more like a sack, intended to maintain the sterility of the operating room) and accompanied his son and me through the double doors and into the operating suites.

A few steps inside the double doors, with another fifty feet or more to the OR, the father finally spoke.

"I think I've had enough."

I sized the situation up; it was clear this man did not wish to be here. I looked at his son, who was totally comfortable, then spoke to the father. The problem was that he had said something too soon. I'm quick, but not this quick. Anesthesia could not possibly be induced in the brief amount of time that had elapsed since we left his wife.

"Sir, one piece of advice. Stand here for sixty seconds, then head back to your wife. If you turn around and walk out right now, it's too quick and she'll know. The switch for the door is on the wall behind you."

"Thanks."

I leaned over Sam and whispered: "Don't tell your mom."

"I won't," he responded. He was quite happy and remained calm throughout the induction of anesthesia.

All went well, and if his mother suspected her husband's failure to be in the company of their son through the induction of anesthesia, she didn't let on.

ANOTHER DAY, ANOTHER ANXIOUS PARENT.
Chase, a two-year-old about to undergo surgery, ran

through the halls of our outpatient center playing with the toys, oblivious to his surroundings, and to his imperfect penis. His parents were not oblivious. For this boy, the question wasn't whether he should be circumcised, but how to fix his original circumcision. By my standards this isn't a significant surgical procedure, but it is a procedure, nonetheless, that requires anesthesia.

A case such as this offers an example of the most routine of the anesthesia care I provide.

An anesthesia resident prepared everything for the case. She approached me and announced: "Doctor Jay, I saw the patient. He is an otherwise healthy two-year-old that has a skin bridge on the penis. He has no other problems. This will be his first anesthetic, and the mom seemed a little nervous. She says she wants to come back with the boy until he is anesthetized. I told the mom you would talk to her. I didn't make any commitments."

"OK," I said. And I thought: *So this is how the day is going to go.* A nervous parent present at the induction of anesthesia in a child never makes my job easier. It's just one more concern for me in the procedure room, one more person to watch, and there is no evidence that parental presence during the induction of anesthesia improves outcomes.

After grabbing a cup of coffee in the lounge, on the way back to the patient intake area I saw Chase happily playing. After reviewing his chart, I talked with his parents. Before approaching the topic of parental presence during anesthesia induction, I discussed every step of the process until he would be anesthetized.

"From the time I walk through this door . . . ," I said, pointing to the door of the pre-anesthesia room, and I recited my two-minute story.

An incredulous look came over the mother. "That's all it takes?"

"Yes."

She stood quietly for a moment, and I was unsure whether I sensed relief or disbelief. She still looked as if she was on the emotional precipice—swaying, ready to tip into meltdown. I tried the best I could to earn her confidence, and I waited for her response.

She turned to her husband, standing quietly beside her. She took a deep breath and then unexpectedly burst out: "Take him. Just take him." Even though I had not addressed the issue of parental presence at the induction of anesthesia, she had come to a decision. To my mind, a good decision.

I scooped up her little son. But after ten steps or so, just beyond the double doors, I felt remorse. Did she really believe me? Had I given her confidence? Or had I coerced her?

Fifteen minutes after separating Chase from his mother, I went to talk with his parents in the waiting room.

"We're done. He's in recovery."

"You're kidding!"

"No. He's really in recovery. We're done."

I don't want to minimize my anesthesia care too much. After all, it does entail risk. There are books many pages long and many pounds heavy that list everything that can go wrong. There are scary case reports in medical journals that document what does go wrong.

A little later, after the parents were reunited with their son, I went to check on them and make sure the boy was comfortable. His mother couldn't thank me enough; then she announced that she hadn't slept all night. She had gotten out of bed in the middle of the night, gone to the bathroom, and sat on the toilet and cried for an hour. Her husband, standing in support behind her, arched his eyebrows and nodded, shaking his head up and down, confirming his wife's story. All I could say in response was: "I wish you had called me. I could have made your experience easier. Have a happy life—and, oh, don't need me anymore."

CHAPTER 6

Nothing by Mouth

———

I T BEGAN IN A DIMLY LIT ROOM WEDGED BETWEEN the postanesthesia care unit and the procedure rooms, a room used as a pre-op holding area for hospitalized patients. The only warmth of this room was a street sign above the door that read "Linda's Place"—white letters on a green background. It was a memorial to the nurse who, for many years, had staffed the room before succumbing to breast cancer far too soon. She used to blush instantly anytime I flashed her a John Belushi grin, a one-raised-eyebrow gaze. I would watch the crimson tide rise up her neck to the top of her head.

In this shadowy space, dwarfed on the adult cart, Michael sat flashing an infectious smile that commanded attention. He was all alone, but unafraid, grinning wide, his teeth fully exposed and made brighter by the darkness of the room. With big brown eyes, bigger and browner than I thought possible, he had the appeal of a teddy bear that you want to give a giant hug. He was just four years old, but beneath his welcoming exterior, a very different, knowing-beyond-his-years boy stopped me in my tracks. Before I was able to say a word, he sized me up, knowing my role as the anesthesiologist—the sleep doctor—and he played his card with panache.

"I had Cap'n Crunch," he said.

"What did you say?"

Silence.

"When?"

Silence.

To many, the broad, inviting smile, bushy white mustache, and wide British naval captain's hat and eyebrows of Cap'n Crunch breakfast cereal bring nothing but morning joy (and plenty of sugar). But to me, they call forth the memory of an anesthetic adverse event—a complication—that I will never forget.

Michael's was my second case of the day. I had worked with his surgeon on the previous case, during which we had discussed Michael's condition and the plan for his scheduled procedure. He suffered the consequences of a congenital bowel obstruction. There are numerous reasons for blockage of a newborn's intestines. A segment of bowel may fail to develop (atresia); or a segment may be twisted shut in what is known as a volvulus or a constricting band or an incarcerated (trapped) hernia; or the nerves may fail to migrate fully to the muscle in the colon wall, leaving it unable to contract and propel its contents forward (Hirschsprung disease); or the opening from the rectum may not form correctly, known as an imperforate anus. Michael's was a complicated case that involved his anus and unfortunately required frequent trips to the OR for exams and dilation. Even at his age, he knew the routine.

His comment about Cap'n Crunch stopped me cold. A stomach filled with cereal and acid—that's how I interpreted his comment. And it had serious consequences for my anesthesia plan.

FIFTEEN MONTHS AFTER THE first public demonstration of surgery under inhalation gas anesthesia, the first death by anesthesia was reported. In 1848, a fifteen-year-old girl named Hannah Greener went to her doctor's office for the removal of an ingrown toenail. She died shortly after breathing the anesthesia gas. The exact cause was not determined, but on autopsy, her lungs were found to be congested with blood and fluid. It's possible the primary cause was heart related—a heart rhythm incompatible with life. It's also possible that she aspirated stomach contents into her trachea and down into her lungs. Very likely, the event caused laryngo-spasm, or "dry drowning."

While consciousness is lost after inhalation of the anesthesia gas, some body reflexes remain intact until the anesthesia level reaches an adequate depth. One such reflex, the laryngeal adductor reflex, causes the vocal cords to close. Before being fully relaxed by the anesthetic, the vocal cords may become irritated and forcefully close. The patient continues to try to breathe against the closed cords, causing injury to the lungs, pulmonary edema, or congestion, and injury or death by asphyxia, lack of oxygen, if not promptly treated.

The discussion of Hannah Greener's death included the observations that her stomach was full of food, and water and brandy were poured into her mouth in an attempt to revive her. Either might have entered her lungs, or possibly she had not been anesthetized to the point of shutting down her laryngeal adductor reflex and her vocal cords had closed tight. Laryngospasm is an anesthesia complication. Whatever triggered the problem, she suffocated.

Nearly a hundred years after the unfortunate Hannah Greener incident, an obstetrician, Curtis Mendelson, observed that women during delivery are prone to aspirating the stomach contents, in part because of the changes of pregnancy—specifically, the uterus full with the fetus pushing everything in the abdomen upward.

The term "Mendelson's syndrome" eventually yielded to the more descriptive "aspiration pneumonitis." The goal prior to anesthesia is to allow enough time to ensure that food exits into the intestines and leaves the stomach empty, eliminating the possibility of aspiration pneumonitis.

The separate paths leading to the stomach and the lungs work on the either/or principle. Only one path may be open at a time, and eating and breathing are kept separate through a series of coordinated actions including muscles and reflexes. The vocal cords at the entrance to the trachea snap shut when food or drink enters the mouth. This reflex, the laryngeal adductor reflex, is beyond our active control and prevents anything in the mouth from going down the wrong pipe. To swallow, sphincters made of muscle in the esophagus relax and the contents of the mouth slide down into the stomach.

Gastroesophageal reflux—commonly referred to as GERD (for "gastroesophageal reflux disease") and also known simply as heartburn—occurs when these sphincters fail and the stomach contents rise in reverse back up the esophagus. Silent aspiration results when the laryngeal adductor reflux fails. Food and drink entering the trachea can block the airway, preventing oxygen in the air from entering the bloodstream. Aspiration also predisposes a person to pneumonia.

Everyone experiences aspiration occasionally by accident. Either excited or impatient, we sometimes fill our mouths with too much food or drink, or too fast, or with a full mouth, and on a startle the contents go down the wrong pipe. Instead of passing down the esophagus into the stomach, the oral contents slide past the vocal cords and into the trachea, the windpipe. The system of protection that keeps the lungs clean and clear has failed. This is aspiration.

Acid is the anesthesiologist's enemy. The stomach is capable of withstanding the acid it produces as an aid in digestion. But other tissue is not immune to this acid. Herein lies the risk of anesthesia, which relaxes muscles and turns off reflexes. The cinched esophageal sphincters relax, allowing the contents of the stomach to flow to the mouth while the laryngeal adductor reflex no longer guards the entrance to the trachea. To prevent aspiration under anesthesia, the stomach must be empty.

Nil per os. "Nothing by mouth." The pre-anesthesia routine in decades gone by was to write and order: "NPO after midnight." With first-scheduled cases in the morning, this worked fine. For afternoon cases, the patient might be left dehydrated. Today, a kinder, gentler sliding timescale is used, depending on the procedure schedule and the type of food or drink. Clear fluids are frequently allowed up to two hours before anesthesia; they don't fill the stomach with acid and might actually aid in passing its contents out and along the intestines. Fatty foods produce the greatest amount of acid, requiring eight hours to clear acid from the stomach.

TWO OF THE FIRST FOUR ingredients listed on the Cap'n Crunch box are sugar, both white and brown. Given the enticement of an inviting figure on the box, placed at child's eye level in the supermarket, and the allure of sweetness, what child wouldn't like Cap'n Crunch? But with Michael preparing for anesthesia, Cap'n Crunch was a potential disaster lying in wait.

I looked into Michael's eyes. I saw a four-year-old who knew that this comment was enough to cancel his case. But was it true? He needed the medical procedure he was there to undergo. I was not about to cancel it without confirmation.

I always approach the evaluation before anesthesia the same way. First I seek out the reason for needing anesthesia and read the patient's present history, past history, and past anesthesia records. Then I perform a physical exam and review lab reports and studies. Finally, I develop the anesthesia and pain relief plan. The anesthesia equipment is prepared and arranged the same way every time. I aim to prevent errors, adverse events, and complications—and to have no need to correct them.

I review the entire chart, from front to back of the

paper chart, from top to bottom of the electronic chart, every time. The demographic sheet listing the patient's name, medical record number, birth date, and address occupies the front of the chart. In pediatrics, a parent stands between anesthesiologist and child, and this parent signs the consent. The bottom of the demographic sheet provides equally important information: spouse, next of kin, parent names.

Before I touched Michael's chart, I scrambled to put things in order. I had heard him and knew what he said, but I remained unconvinced that he was being truthful. Could a child this age knowingly, actively develop such a lie?

A patient not optimized for anesthesia presents increased risk of complications. Food in the stomach at the time of anesthesia poses one such risk. Anesthetizing any patient not optimized for the procedure practically guarantees an encounter with the "retrospectoscope"— the hindsight that displays so clearly later what couldn't be seen at the time, the "instrument" that raises the question "Why didn't you . . . ?"

"Iatrogenic"—derived from the Greek *iatros* ("physician") and *genic* ("to be produced by")—is a word that was developed less than a century ago to indicate something caused by a physician that has adverse effects. Anesthesia can only be iatrogenic. That is, only rarely

are my efforts meant to treat or cure a patient's condition, such as the occasional nerve block for a pain syndrome gone wild. In the vast majority of cases, my
intention is that a patient whom I place into a state of
unawareness and painless comfort (an induced coma)
will emerge from that anesthetic coma in better health
by my hands. Some responsibility is borne by the patient
and family to follow instructions and adequately prepare for anesthesia. Ultimately, though, the responsibility lies with me.

I looked around the area. No parents and no nurse.
Even the transporter who had brought Michael to me
was nowhere to be seen. With no one to ask to disprove
or confirm Michael's Cap'n Crunch claim, I turned once
again to my little patient.

"Where did you get the Cap'n Crunch from?"

The grin remained frozen on his face. His teeth never
parted, and he said nothing.

"When did you have it?" I prodded again, but still he
did not speak. He adopted an impish look. Those brown
eyes were no longer as large as they first seemed. He
stared right at me, knowing full well the implications of
his comment.

The easy way out was to postpone his case. But he
needed an exam and would be no better without it, so I
proceeded to investigate the Cap'n Crunch claim.

I knew Michael was experienced with anesthesia and surgery. He knew the routine. I suspected he was attempting to trick me into altering my care.

After the demographic sheet and the consent for a procedure, the patient orders follow. Listed clearly was "NPO after midnight."

"The chart says you were given no food today," I said. Again, I got no response, as his smile evolved into a smirk.

Slightly unnerved, I decided to check further. I stepped out of the room and phoned the hospital unit that Michael had come from and spoke directly to his nurse.

"No, doctor. He had no cereal. I made sure his breakfast tray was not delivered today."

I returned to the boy.

"I spoke with your nurse. She just told me you had nothing to eat today."

He broke his silence with a rapid burst of: "I ate Cap'n Crunch!"

"Where did you get it?"—my voice a bit louder and more imploring than before.

"My mom." Then the same grin, the same locked lips, and the silence resumed.

"When did your mother give you the cereal? Your mother isn't here." I received the silent treatment again.

Back to the phone, back to the nurse from the unit. "No, doctor. His mother wasn't here today. It has been a couple of days since she has visited." (A sad, all-too-frequent reality in my practice.) "He didn't have any food today."

I looked at the procedure consent and found that it had been obtained by phone. His mother had not been present the previous day.

After challenging the boy one more time, and receiving nothing but silence in return, I returned to the phone. I had reached the limit of the nurse's patience and had begun to perturb her. "No, doctor," the word "doctor" emphasized and prolonged. "He has had nothing to eat today."

I noted in the chart the lengths to which I had gone to verify the cereal story. I considered whether, if he had indeed eaten the cereal, I should wait until his stomach would be cleared. But I didn't know what time he would have had the cereal. I decided that, at four years of age and with his lack of cooperation, he would not, even if he could, accurately report what time he'd eaten the Cap'n Crunch. An eight-hour wait would deprive him of fluids and possibly cause dehydration. Either he'd eaten the cereal and the case should be canceled, or he hadn't eaten any and I should proceed.

I believed the nurse. Through due diligence, I concluded there was no Cap'n Crunch. Four false words from a four-year-old were not going to dictate my care.

Off to the operating room we went.

After anesthesia was induced, the "urp" was subtle.

Just a tiny flicker of his abdomen. No one except me noticed it. But removing the mask revealed a mouthful of Cap'n Crunch. Yes, he had indeed eaten the cereal.

The next few minutes were nerve-racking. I turned him on his side to prevent anything from entering his airway. I suctioned the cereal out of his mouth. Then, with trepidation, I listened to his lungs with my stethoscope. His breath sounds were clear. He remained pink; his oxygen saturation remained normal. There but for the grace of God go I; he did not aspirate the contents of his stomach, and he emerged none the worse for wear.

Michael later confided that he had sneak-eaten his roommate's breakfast. I resisted the temptation to call the floor nurse.

THE WEIGHT OF RESPONSIBILITY, or maybe the sense of guilt, from this almost adverse event, this near complication, hasn't diminished with time. The colors of the room, the green on the street sign, the time on the clock on the wall, the words exchanged remain just as crisp within my memory as the day it all took place. The drive for perfection persists, and that may be the most damning part of this case. No matter how I reevaluate the facts, my decisions never change.

Given an identical situation today, my care would be the same.

Danger lurks.

I do not mean to incite fear of anesthesia. This four-year-old might have died because of my care, and if he had, he would have been the first and only patient without beforehand identifiable risk of death to do so in my career. It is more probable to be struck by lightning than to die while under anesthesia. The risk of dying from anesthesia is less than one in a hundred thousand cases. It's about the same risk as dying as a result of sky-diving, participating in a triathlon, or riding a bike. Anesthesia is very safe. But anesthesiologists can never, ever let their guard down.

A patient's "physical status" is a numerical assessment by the anesthesiologist of the patient's overall health. A PS1 patient is healthy. A PS2 has a compensated health problem not impacting everyday life—high blood pressure controlled with medications, for example. Eighteen holes of golf presents no problem. A PS3 suffers from a life-altering condition, such as heart disease that makes it difficult to function normally—for instance, to walk up a flight of stairs. PS4 and PS5 patients are at risk of dying or are expected to die. Anesthesia risk rises as the PS number does. It's mildly curious to me that no patient has ever asked for my PS assessment.

In my career of thirty thousand anesthesia cases, give

or take a few, during my care PS1 and PS2 patients have come in healthy and gone out healthy—no exceptions. I have also had patients living between the proverbial rock and a hard place, patients for whom the only chance at continuing life was the procedure room. Not all have survived. I carry, and always will carry, the memory of every one of those patients.

The simple fact is that a healthy, nourished, hydrated patient presents the best chance for an optimal outcome. No solid foods for eight hours before anesthesia. Avoid fatty foods. Only clear fluids, fluids you can see through that don't contain fat (in other words, no soups and broths) until two hours before the scheduled anesthesia. Control what can be controlled: blood pressure (hypertension), blood glucose (diabetes), and airway disease (asthma and chronic obstructive pulmonary disease, more commonly referred to as "COPD"); if quitting smoking isn't a possibility, at least stop for several days. Put the data in your favor.

Although death during anesthesia is rare, adverse events do arise. Postanesthesia nausea and vomiting is the most likely issue. Placing a number on the incidence of nausea and vomiting after anesthesia is difficult because nausea is a subjective complaint, and even vomiting is not so easy to evaluate. Spitting out oral secretions is not vomiting. Nausea and vomiting are also procedure specific. Eye surgery is notorious for postprocedure nausea and vomiting (PONV), with reported rates of up to seventy-five percent. The often-quoted incidence of PONV after general anesthesia is thirty percent. In my experience, fifteen percent is more accurate. I surveyed my cases over a several-month period and discovered a six percent incidence of PONV in the first twenty-four hours after general anesthesia.

Damage to teeth is another of the more common complaints. Again, the numbers vary widely, from less than one percent up to six percent.

These two issues are what I term nuisance problems. I don't mean to minimize their impact, but these issues don't risk life. Of the more common and real adverse anesthesia events, respiratory complications occur at a rate of about one to two percent. Respiratory complications range from less-than-expected oxygen levels (low pulse oximeter readings) to pneumonias and aspiration pneumonitis.

ADVERSE EVENTS BEYOND THOSE of a respiratory nature fall into the "rare" category.

Early on in medical school, a lecturing doctor introduced a woman to my neurosciences class and asked her to flap like a bird. By the fifth flap she could hardly raise her arms. We were witnessing the effects of a disease named myasthenia gravis (from the Latin for "muscle weakness" and "grave" or "heavy"). In this condition, the receptor on a muscle that receives the chemical transmitter released at a nerve ending as an instruction prompting a muscle contraction doesn't work. The condition is very rare. Even in my high-risk practice, I see only about one case of it each year.

The anesthesia equivalent to myasthenia gravis is frequently termed an "allergy," but it is not. A reaction to certain anesthesia drugs leads the muscles of the body to run amok, resulting in high fever, high heart rate, and high carbon dioxide levels. If unchecked, it leads to death. Malignant hyperthermia (MH) is a genetic variant that predisposes certain patients to this metabolism-gone-awry reaction, but the condition is extremely uncommon. An anesthesiologist with a busy practice of both adults and children will encounter one case of MH every thirty-seven years. Although as similar in incidence to myasthenia gravis, every anesthesiologist is required to know all there is to know about MH, and essentially no one should die from it.

The potential risks of anesthesia—both common and rare—require anesthesiologists to remain vigilant and to maintain a wide scope of knowledge to ensure optimal outcomes. Every box of Cap'n Crunch serves as a reminder that anesthesia complications lurk in the shadows.

CHAPTER 7

Heartbeats

———

BEFORE I ENTERED MEDICAL SCHOOL, I NEVER dreamed that the sight of the human heart, exposed and beating in front of my eyes, would move me so deeply. The first time I saw it was a religious experience. The brain is the most complex organ, but its energy isn't visible; there is no action, no movement, and thoughts can't be seen. The heart is the most inspirational organ to observe, to touch, to feel. Nothing stirs my medical emotion more than watching the heart deep in the chest, lurching with every beat. Unlike any other organ, it's always in motion, never resting. Its beat defines life.

The size of a fist, the heart, with its maroon surface, obscures the red in a chamber lying beneath. With patches of pale-yellow fat adding to its shiny surface, its many curves and rhythmic movement reflect the light of surgery in a continuously moving array. In the heart's vault of ivory-colored walls, in a shallow lubricating bath, its every movement creates a sloshing sound, "la-lup," a couplet unmistakable and instantly identifiable.

The surface of the heart is warm, moist, a little slimy. It brings to mind the sticky mucous covering of a night crawler, like a thin coating of oil. Transformative energy flows from its imperceptible electrical charge, providing a spiritual awakening for the exploring finger. A reflexive jump instigated by touch results in a momentary stillness; then its constant rhythm— "la-lup"—quickly resumes.

Through a stethoscope, the couplet sounds crisp, with snaps as the valves inside open and close in synchrony. The path of electrical energy originates in a small clump of cells coursing through the heart, flowing along cables and morphing into a neon green line across a flat-screen monitor: a straight, horizontal section broken with a small hill, then a series of jagged up and down peaks followed by a larger hill as the electrical charge moves down the heart. After a quick rest, the landscape repeats. This whole complex creates a single beat of the heart.

My identity as a physician took form as I wandered into the field of anesthesiology. But it was not completed until I found my passion in working with children—and the heart. And so I evolved into caring for children with sick or deformed hearts.

JOHN WAS A HIGH SCHOOL sophomore and dreamed of playing basketball. He loved the deep, resonant thud of leather bouncing off wood and echoing through the hall; the screech of athletic shoes on the floor; the sight of the ball sailing through the air, its seams revolving like a world of its own; and the sound it made passing through the rim—nothing but net, that distinct swish—before falling to the floor. He was about to get his shot—again. By his mother's account, John had been the last player cut from the team his freshman year. Now, a year later, with hard work and determination, he would realize his dream of making the team. After all, Michael Jordan had been cut from his high school team too.

Tryouts turned into a nightmare before warm-ups were completed. John's body couldn't deliver on the enthusiasm of his mind. His legs failed to perform to the level of his determination. Running up and down the court, John lagged. Another player dribbled the ball right around him. John's legs felt leaden. Lacking speed or energy, he was losing hold of his dream, and he knew it. His anxiety grew as, the harder he tried, the more he gasped for air. Something was wrong.

He told his parents about his trouble and soon found himself speaking with his pediatrician, who referred him to a pediatric heart specialist, who referred him for cardiac catheterization to determine the cause of his failing energy. I provided his anesthesia care when he presented for a diagnostic procedure.

I spoke with John and his mother before the procedure, which would require a trip to the cardiac cath lab. After examining him, I sensed that his basketball season was finished. Indeed, my concern was no longer for this season, or the next, but for whether John would ever play basketball again.

He rested comfortably on the cart, and his health history was unremarkable, aside from the fatigue and shortness of breath. On physical exam, John was a little chunkier than I expected for a basketball wannabe. His belly was fuller than anticipated. This wasn't fat; he was overweight from fluid retention. His heart, unable to keep up with the returning blood, was causing his liver to engorge, pushing his belly out. Excess fluid filled his belly and caused his legs to swell. Not dramatically, but any swelling was problematic. I came to my own diagnosis but didn't mention it. It was neither definitive nor mine to make, so I limited my discussion to the intended anesthesia care. But it was clear to me that John's heart was failing and he was in trouble.

John's proposed trip to the cardiac cath lab was not

intended to provide therapy. There were no clogged arteries needing to be widened with a stent, no incompetent or narrowed heart valves needing to be repaired or replaced, no holes between chambers needing to be shut. This procedure was meant to provide nothing beyond a diagnosis, and the cath lab was the diagnostic tool of choice.

The cardiac cath lab is a cold place. It is chaos on a tether. I have entered this room hundreds of times and always find little evidence of cohesion. The patient entrance opens to an expanse of vacant space lacking a discernible purpose. Halfway across the room is a floor-to-ceiling mass of technology that appears unnecessarily crowded. On the inverted maze of runways that constitutes the ceiling, a series of beams, bars, pipes, and cables crisscross above an anchored procedure table. From the beams drop the booms supporting large fluoroscopic—X-ray—tubes that rest to either side and above and below the procedure table. A row of flat-screen monitors, in constant territorial conflict with the fluoro units, hang from the ceiling.

Every piece of equipment fights to be the closest to the procedure table: the anesthesia machine, attached gas lines, monitors with electrical cords, an anesthesia supply cart, a defibrillator cart (to deliver an electric shock to an erratically beating heart), an ultrasound machine that uses sound waves to spy harmlessly inside the body, and a multipurpose cath cart. Around the table are the poles holding IVs and infusion pumps that supply drugs to the patient. The cords and tubes coursing from the procedure table present an unfortunate risk for tripping.

This place is a danger to head and foot. In this room, if I collapsed in my anesthesia nest, there would not be enough space for me to land on the ground.

Above the maze of ceiling structures sit can lights whose rays, in a failed attempt to light the room, are broken by the vast array of suspended equipment and instead cast shadows in all directions. The patient on the table can easily seem lost in this jungle of technology. Murals rest high on the walls near the ceiling, attempting to soften the room with the image of a biplane soaring through white, billowy clouds while a hot-air balloon and a kite float nearby. But the view is obstructed, and the patient would need to strain to see it.

WHEN JOHN'S GURNEY WAS PUSHED to the procedure table, his hand cautiously fingered the edge of the table, and he began to slowly transfer himself. As he moved, he needed to avoid hitting his head on the fluoro unit hovering within striking range above the table. The effort he exerted confirmed my belief that his heart was failing. I noted his energy deficit as he moved. I needed to consider the extent of his heart disease, and whether it was more involved than I believed, to determine whether my plan for anesthesia should be revised, and whether my intended drugs needed to provide more protection for his heart.

A failing heart and the induction of anesthesia are diametrically opposed. Drugs used to induce anesthesia depress the heart and can tip a patient like John over the edge. In an anesthetized state, his heart might not be able to beat strongly enough to provide adequate blood flow or pressure. That is, John could go into shock or suffer cardiac arrest. John presented me with the precarious job of striking the right balance between providing comfort and maintaining sufficient function in an already underperforming heart.

I considered making changes to my plan to prevent drops in blood pressure or heart rate. But for every pro there is a con, and my second-choice drug—the drug I would select if I thought his heart was critically failing—can cause unpleasant hallucinations.

When John centered himself on the procedure table and lay flat, he appeared comfortable. My mind eased because a poorly functioning heart would not allow such a level of comfort. His breathing would be too labored as his liver pushed on his lungs, already congested with excess blood. Rather than being anxious, I was vigilant. (An anxious anesthesiologist is prone to making poor decisions.) With the drugs from my original plan, John safely entered the state of anesthesia.

After the preparations for the procedure were completed, I stepped back from the table to look at John, lost under the blue sterile procedure sheets. A hole in the sheets maybe five inches across exposed his right groin, giving access to his blood vessels. It was the only patch of skin in sight, the only clue that a teenager was actually present. I was connected to John solely by the sounds from the monitor calling out his heart rate and oxygenation, with the lines of his well-being crossing the monitor screens. It was easy to lose sight of the fact that this was more than a sick heart—that a whole life lay in the balance.

A biopsy forceps, a wire several feet in length with the tip holding two tiny articulated cups that snap shut and grab a piece of heart not much larger than the head of a pin, entered John's femoral vein through a needle and catheter placed in his right groin, then snaked up and into his heart. Little chunks of heart muscle were then snipped from the inside of John's right ventricle, the pumping chamber to his lungs.

Lymphocyte infiltration, myocyte necrosis, endocardial fibrosis. Microscopic examination confirmed the cause of John's heart failure. In plain terms, his heart muscle was inflamed, dying, and being replaced by scar tissue. His condition was most likely the result of a viral infection, a myocarditis. Scar cells don't beat and are unable to pump.

After the procedure, John's parents' reaction when told of their son's condition was predictable: "How do we treat this?"

They stood stunned and muted by the response: "A heart transplant."

Just weeks before, John had been a normal high school sophomore unaware of any health problems, preparing to star in basketball. Now he was dying.

AT THE MOMENT OF his diagnosis, John and his family entered the dark side of medicine. In a twisted sense of the balance of life, for John to live, someone had to die. A viable, undamaged organ for transplant requires a quick strike to life, causing rapid brain death. The scenario almost always involves violence. The optimist views it as making the best of a horrible situation— saving one life instead of losing two. Still, the surgical team that harvests the organ, a team from the hospital that will implant the organ into another person's chest, reminds me of grave robbers from centuries ago, who, with shovels in hand, looted cemeteries during the middle of the night by the light of oil lamps. Today, they are looting a body in a bright, clean, and sterile OR.

The downhill spiral of John's failing heart was rapid. After his cardiac cath, he remained in the hospital and was placed on inotropic support—meaning that he was given intravenous medications to boost his heart's ability to contract—while he awaited someone else's misfortune. Fortunately for John, and unfortunately for some other poor soul, the progress toward his heart transplant was a sprint. Several weeks after his diagnosis, someone died suddenly, and that person's heart came to beat beneath John's breastbone.

During his heart transplant, looking over the ether screen into the chest cavity (with the paper drape, known among anesthesiologists as the "blood-brain barrier," dividing the surgical area from my anesthesia nest), I saw that John's heart wasn't the size of a fist, but more like a melon, a cantaloupe; the firmness of a healthy muscle had given way to the bagginess of failure. And rather than bouncing, it was rocking from side to side, with the "la-lup" sound faint and indistinct.

After the transplant, which was successful, John received anesthesia on multiple occasions, including for subsequent heart caths for biopsies to measure his body's attempt to reject his new heart. His initial recovery was remarkable for being unremarkable. My concern that anesthetics might hurt his heart lessened. He now contained a healthy heart.

Recovery for anyone receiving a heart transplant is a

never-ending marathon. John's new goal in life was to maintain the health of the heart in a foreign home, an unwilling and hostile environment that would rather not house it. He had to take multiple medications to stop his body from attacking the foreign heart muscle.

These drugs must be taken on schedule. Maintaining constant drug levels in the blood is imperative because the body's defense mechanism for fighting invaders, in this case the foreign heart muscle, never rests. There are no medication holidays.

The standard seasonal illnesses that cause the common cold and the stomach flu become life threatening for the transplant patient if medications can't be swallowed or held down. Hospitalizations are repeated and necessary to provide intravenous fluids and medications on a timely basis.

The longest-surviving transplanted heart to date has beat a little over a billion times, or thirty years. Eight out of ten heart transplant recipients survive the first year. Most will survive for twenty years, or 750 million heartbeats. That's a very large number, but it is far less than three billion, the number of heartbeats in a normal life. So, the transplanted heart is not a guarantee into old age. It is a lease on life that will need to be renewed. This is especially true for young recipients. The transplanted heart gradually develops narrowed blood vessels—a vasculopathy—a perverse injustice resulting from the same medications that prevent organ rejection. Heart failure eventually recurs.

John's parents dedicated themselves to his welfare. His medications were always available, his schedule maintained, his checkups routine, all recommendations followed. For teenagers, this rigid schedule is difficult. It's a confinement that ostracizes these kids in times of rapid growth—physically, psychologically, and socially. School trips, outings with friends, late-night parties, and vacations all harbor threats. Drinking binges and hangovers are forbidden. Life is never as it was, or should be.

But John was committed to the regimen, and a couple of years later, he outgrew our pediatric hospital and his care was transferred to adult medicine.

AROUND THE SAME TIME, eleven-year-old Bandul came to us with his heart in worse condition than John's. The son of undocumented immigrants, he was also uninsured. Few countries in the world would allow Bandul, unable to cover his costs, to enter an ER, let alone provide all necessary care until complete resolution of his heart disease.

With a bucket of unexpected surplus revenue, a bonus from an uptick in the economy, the state legislature had opted to pass the good fortune on to sick and needy kids. A Medicaid plan (state-managed care for the poor), increasing funds for children of families unable to afford health care—regardless of citizenship or even immigrant status—found receptive legislators not worried about raising taxes and losing votes. Bandul benefited from this new plan and received the same quality and quantity of care as John, no questions asked. At least from my anesthesia perspective, the care for Bandul and John was identical.

But after Bandul's cardiac cath and diagnosis, he navigated a more ominous course than John's. Bandul's heart was failing more dramatically, and it worsened during the wait for that unknown someone special to die. He required a procedure attaching him to an external pump that assisted his heart's contractions and blood flow—a ventricular assist device (VAD). Not only was his anesthesia complex, but just moving him from the ICU to the OR was a monumental task filled with potential risk. Luck shone on Bandul, though, as he, too, received another unfortunate's heart.

After the transplant, Bandul remained in the hospital longer than John did. His native heart had been unable to support his kidneys, causing damage that took longer to heal after his heart transplant. But Bandul was eventually discharged on the same medications, and with the same needs, as John.

Great lengths were taken to ensure that Bandul's

non-English-speaking family understood the need for the medicines to be taken on a timely basis, for return visits to clinics not be missed. (On any given day, twenty percent of Medicaid outpatient appointments are missed.) Any variation from the rigid schedule could injure Bandul's new heart.

Bandul's parents were diligent, and he returned as planned for all of his appointments. A strong support group was built around him. As Bandul grew, the tone of the physician notes changed. At eighteen, Bandul was charged with the responsibility of taking his medications unprompted and of scheduling his own appointments. Bandul's chart indicated that on questioning, he knew all his medications by name, could list the times for taking all the pills, and knew the phone numbers for his doctors, clinic, and hospital by heart.

But death always lurks near heart transplant patients.

I met Bandul again when he was midway through his twentieth year. The night before, he'd been admitted through the ER with complaints of abdominal pain and diarrhea. He needed a plastic tube threaded through a vein in his arm to his replaced heart to administer his medications and obtain blood samples. Bandul's body was rejecting his transplanted heart.

He appeared healthy—thin but robust. He was buff, with firm biceps and leg muscles. And he was polite, readily answering my questions.

"Why did you get that tattoo?" I asked, referring to the giant tiger tattoo overlying his biceps on one arm. That tattoo in a patient with a transplant indicated risky behavior and rash decisions.

"It's for kickboxing."

"What?"

"It represents fighters and kickboxing."

"You kickbox? You've had a heart transplant."

"The doctors said I could. They said it was all right if I was careful."

Bandul's anesthesia was uneventful. His heart rate and blood pressure formed railroad tracks on the anesthesia record, the straight and flat lines of X's and dots that I crave. After the procedure, and without apparent harm, I brought Bandul back to the intensive care unit.

But Bandul returned the following week because fluid was building up around his lungs, in his chest cavity—a pleural effusion. A tube to drain the fluid was needed. Bandul was in a torrid downward spiral. In medical lingo, he was circling the drain.

Another colleague was scheduled for the case, but because Bandul's condition was tenuous, a pediatric cardiac anesthesiologist was required, so I met Bandul once more. I again reviewed his chart prior to the procedure, and the issues associated with his hospitalization were recorded more completely.

My own heart once again felt like a lead weight was

hooked to it, pulling it down and out of my chest and into my belly.

Several notes by physicians and social workers indicated that Bandul had failed to take an antirejection medication for nine days prior to coming to the ER. This anesthesia record was not railroad tracks. There were a few bumps. His blood pressure varied—not severely, but noticeably—and Bandul was placed on medications to support his transplanted heart's function.

My interactions with Bandul were always pleasant and polite. When I transferred him from the cardiac cath lab to his room, his parents—as devoted as could be—sat on a couch beyond Bandul's bed and leaned forward in silent anticipation, his father with his hands clasped together between his knees, his mother with her hands folded in her lap. Their eyes grew large and forlorn, as if they knew the extent of Bandul's illness, praying that I held the answer, the cure.

In addition to having me as the pediatric cardiac anesthesiologist, Bandul received care from the ER physician, the pediatric transplant cardiologist, the pediatric interventional cardiologist, the pediatric interventional radiologist, the pediatric pathologist, the congenital cardiac surgeon, and the pediatric cardiac intensivist (critical-care physician). And then there are the nurses and all the other technicians necessary to complete this team of astounding expertise. But despite the mastery of the team, our reach beyond the walls of our medical center is not all-encompassing.

Teenage brains lag behind their bodies in development. A kid may look like a sleek and buffed sports car on the outside, while what's under the hood isn't yet perfected. Teens have pedal-to-the-metal accelerators, but they lack competent brakes. Subtle changes to the brain continue into the twenties, gradually containing the twitchy impulse control—or lack thereof—and the emotional volatility of the teen years.

The law expanding Medicaid coverage and providing care to all children, no questions asked, resulted in an unintended consequence. The benefit wasn't never-ending. The gratuity needed a sunset. While physicians as a whole, and especially pediatric specialists, focus on a patient's life for decades ahead, no end of care exists. The writers of the law failed to recognize the needs of a patient like Bandul eight years after his transplant.

The politicians decided that their generosity ended when kids were no longer kids. They defined that moment as the last day of the month of a child's nineteenth birthday. So, Bandul was no longer provided his medications. Bandul worked with his father at odd jobs as they became available, and he dreamed of attending college and entering the medical field. The school he inquired about was a drive away. On impulse, he chose savings over purchasing the transplant rejection–preventing medication that now cost him twenty-five dollars per day. Just the first few moments of Bandul's prolonged intensive care stay consumed the two hundred dollars he had saved, as the hospital room charges alone came to hundreds of thousands of dollars. The total of the charges accumulated during his stay, including tests, therapy, and procedures, would have paid for years of Bandul's preventive medications.

Bandul died on hospital day fifty-seven.

I play no role in the decisions made beyond my involvement in a case. My role and my goal are simply to provide the most skilled anesthesiology care possible. And I don't function alone; I am a proud member of one of the elite teams in the world of medicine. But the success of the team, and the health of the child heart recipient, often lie outside the team's grasp.

CHAPTER 8

A Most Unusual Patient

D URING MY SURGICAL TRAINING, A VERY GIFTED surgeon offered a surprising piece of advice: "Treat every patient like dirt. Then every patient will be treated exactly the same." His intention was clear: treat everyone as equal so that a patient's status doesn't alter the process. My take was to treat everyone the same, but like royalty.

I always remind myself that there are no VIP patients. But some patients are more unusual than others.

After a three-day holiday weekend, I walked into the hospital and approached the surgical control desk. When I'd left the previous Friday, I hadn't yet been assigned any cases for the following Tuesday. Over a weekend—especially a long weekend—schedules change.

There is plenty of time for patients to be admitted, for surgical illness to arise, for injuries to occur.

The fellow in pediatric surgery (fully trained in general surgery, now training in his subspecialty) stopped me short of reaching the desk.

"You free?"

"I don't think I'm on the schedule. Why?"

"Come with me. I need a hand."

Retracing my path down the hallway and out of the hospital, we unexpectedly turned into the research building. This building housed the facilities for virtually every type of research, from running chemical reactions in test tubes, to growing cells, to the rare larger animal experiment. I didn't think the surgical fellow was asking for help on a chemistry or microscopic cell–based project. Although trained as a chemist, I hadn't practiced chemistry for many years, and I had never attempted cell-based research. My imagination began to run.

Health care research is a necessary minefield. It's a bit of a bet: asking the right question, developing a plausible solution, spending time, energy, and money collecting data—all in the hope of determining a definite answer. A long road to potentially improved care and outcome lies distantly ahead. No alternative to research conducted on the living yet exists, and we probably won't ever be able to resolve all the ethical issues involved in such research.

Research studies on humans are often attached to required care. Patients suffering an illness are offered alternative experimental therapies, trial drugs, or untested techniques in addition to the accepted drugs and procedures used to treat their condition. For example, a patient undergoing knee replacement surgery might be asked to take part in a study on postoperative pain relief, but the surgery will take place whether or not the patient participates in the research protocol. Human studies have opt-out clauses. An unsatisfied patient, or one who has a change of heart, can voluntarily end participation in any study at any time.

Animal studies are a very different story. Healthy animals tend to be used in research, and it's hard to do no harm when nothing is wrong to begin with. The onus rests on the researcher to demonstrate that the research is warranted, that an animal model is a necessity, and that the animal will not be hurt or suffer throughout the study. Animals are unable to voice their discomfort or invoke an opt-out clause. As a result, the demands on researchers to ensure the animals' well-being tend to be more stringent than for human studies, and there is zero tolerance regarding any potential for the animal subjects to experience pain.

I've consulted on animal care protocols, ensuring that no breaks in the standard of care exist, that comfort is complete. But just as animal anesthetists are not versed in human anesthesia, I am not versed in animal anesthesia. When I wrote and submitted a research proposal for an animal study that would investigate how cardiopulmonary bypass affects lung function, it was initially rejected for not meeting animal anesthesia standards as determined by veterinarians and the facility's institutional review board, even though my proposed anesthesia exceeded the standard human requirements accepted by anesthesiologists.

ON THE WALK TO the research center that Tuesday morning, as my mind wandered over the possibilities, I was excited to think that my expertise might be needed to address a research issue.

Entering the animal procedure room, I froze. My expertise hadn't been requested for a research dilemma. In the center of the room, on an obsolete human OR table repurposed for research, a gorilla lay motionless. She was named Tabibu. Even as sick as she was, Tabibu was stunningly beautiful. Dwarfed by the adult-sized OR table, at less than two years old, she was about three feet tall and perhaps thirty pounds, with intensely dark-brown fur and skin a deep midnight black. I stepped to the bed, and her two handlers shuffled away from her, perhaps accepting the gray of my lab coat as authority. Tabibu's open eyes were the color of dark-roasted coffee, but they were dull and didn't reflect my image.

I melted. Before I could kick into detached physician mode, I needed to conquer my empathy—that hug-the-puppy desire. I listened as the surgical fellow and one of the animal handlers, who turned out to be the zoo vet, provided me with the details of Tabibu's condition.

Three days earlier, my anesthesiologist colleague on call for that weekend's emergencies, Andy Roth, had received a call from our pediatric surgeon, who in turn had received a call from the zoo. Late the week before, Tabibu had become severely ill with an acute abdomen. In humans, an acute abdomen is severe—an incapacitating belly pain that develops over a short period of time. The patient is bent over in agony, unable to stand straight. Lying flat on a bed, the patient avoids all movement because moving causes the abdominal contents to shift, triggering a pain response.

The zookeepers had noticed that Tabibu had stopped eating, become lethargic, and retreated from all social contact. Concerned, on Saturday morning Tabibu's keeper had called the zoo vet. In comparison to human medical specialties, the zoo vet is the family doc caring for hundreds of species—from snakes to birds, pygmy shrews to rhinoceroses. The zoo vet's net of knowledge is cast wide, but not necessarily deep.

This vet understood the severity of Tabibu's illness

and realized that the situation exceeded her ability. She contacted our pediatric surgeon because a juvenile great ape physiologically and anatomically is not too dissimilar from a small child, and a gorilla's genetic profile is at most a couple percentage points different from that of a human. The surgeon recommended transferring Tabibu to our research facility for evaluation. Our surgeon confirmed the diagnosis of acute abdomen. Tabibu needed surgery. Her abdomen might contain an infection, an obstruction of the intestines, or a disruption of blood flow (ischemia) to an organ.

Andy was asked to provide anesthesia for an exploratory laparotomy. In this procedure, the surgeon opens the patient's abdomen, exposing the organs to determine what's wrong. This is a major procedure in elective situations, and in emergencies it is fraught with risk, both from the anesthesia and from the surgery.

Once Andy Roth had safely anesthetized Tabibu, surgery revealed an infection that was eating away part of the wall of her colon. The diseased section was removed, and she emerged from anesthesia as hoped.

A human patient who undergoes an abdominal surgery will not be released from the hospital until clearly able to eat and drink, and after having a bowel movement. Concerned for the safety of both Tabibu and her keepers, however, the vet insisted that Tabibu return immediately to the zoo. But the vet had not taken into account the impact of the surgery on the patient. Back at the zoo, Tabibu's problem wasn't pain, which could be—and was—controlled by her keepers, but the fact that surgery on the bowels results in the condition of ileus: her bowels stop working. Muscle in the walls of the intestines stops squishing and propelling the contents downstream. Fluids and nutrients were not being absorbed.

A chain reaction results from not eating or drinking, and dehydration sets in if intravenous fluids are not supplied. The heart rate rises in an attempt to maintain adequate blood flow, but eventually the blood volume shrinks beyond the point of compensation by the heart and the blood pressure falls. This is what doctors refer to as shock, and it can be fatal if not promptly reversed.

During my care, I held Tabibu's hand and rubbed her head. I grasped her arm; it seemed thinner than a child's as my fingers encircled it. Her fur felt stiff and wiry and her skin thick. The only movement she was able to make was to roll her upper lip out. Her inner lip was dry—another sign of dehydration. She didn't withdraw or resist my advance in any way, a not-so-subtle indicator of how ill she was. Her breathing was what a physician would describe as agonal: rapid and shallow, with an occasional sigh and pause.

Tabibu's heartbeat was fast, but her EKG appeared

normal, meaning that her heart was beating with a regular rhythm. That was a good sign. However, it was
clear to me—from her heart rate and breathing—that
Tabibu was in shock and was approaching cardiac arrest.

THE DOSE OF THE ANESTHESIA gas is measured as a
percentage of the overall inhaled gases (as mixed with
air and oxygen). The minimum alveolar concentration
(MAC) is the percentage of gas inhaled that prevents
fifty percent of patients from responding to a painful
stimulus. Whether it's a mouse, red-tailed hawk, monitor lizard, elephant, or human, regardless of species or
size, the percentage of inhaled gas necessary to achieve
the state of chemical coma is remarkably similar.
Greater change exists with advancing age than between
species: once you hit maturity, the older you are, the
less gas you need.

The same cannot be said for the anesthesia drugs
administered by injection. Differences in species alter
the IV (intravenous) or IM (intramuscular) dose of the
drugs necessary to provide anesthesia. More to the
point, the required amount of drug increases as the level
of oxygen consumed increases. Small species tend to
consume oxygen in amounts that are magnitudes higher
per pound or kilogram than the amounts that larger
species consume and, as a result, anesthetizing small

species requires larger doses of drug. The dose per pound of an anesthesia drug injected into a human might kill an elephant but leave a mouse unfazed and staring at you, wondering what just happened.

Providing anesthesia for Tabibu was, intriguingly, not that different from caring for any eighteen-month-old child. I prepared my space around Tabibu's head. I checked to ensure that I had the ability to provide oxygen and to breathe for Tabibu as necessary. I checked my airway equipment—the laryngoscope and endotracheal tube of appropriate size—and prepared my emergency drugs, including epinephrine if her heart needed a kick.

By my assessment of her breathing, Tabibu needed to be intubated, and quickly. She might stop breathing—apnea—at any moment. I chose to use an injectable anesthetic, ketamine, a drug that doesn't cause a change in breathing or in blood pressure, either of which might be fatal for Tabibu. She barely flinched before succumbing to its actions. When her eyes bounced from side to side (nystagmus, a sign that the drug was working), and with her breathing in oxygen-enriched air unchanged, I moved to her side to search for a vein where I could place an IV to replenish the fluid volume. There was no hope of seeing the hue of blue that would indicate a vein beneath her thick, brown skin, but on her right forearm, I noticed the slightest bulge.

I reached for an IV catheter consisting of a hollow

plastic catheter sheathed over a very sharp stainless-steel needle an inch or more in length. The needle enters the vein, and the catheter slides off the needle, coming to rest inside the vein. IV catheters come in a variety of diameters (called "gauges") designated by the color of the hub of the catheter. The larger the diameter, the more quickly IV fluid flows in. I grabbed one with a blue hub (a twenty-two-gauge catheter) and envisioned myself placing that needle into the vein, sliding the catheter off, and feeling it come to rest within the vein lumen.

Frankly, I was being a wuss. Tabibu needed fluid, and lots of it, and a twenty-two-gauge catheter was too narrow to rush intravenous fluids into her body. Considering the amount of fluid she needed, a larger, twenty-gauge catheter (pink by color) would have made me happier, but I lacked the confidence to accurately place it. With only one vein visible, I had one shot, and failing was not an option. Once more invoking John Hughes with "Our Lady of Victory, pray for us," I pierced her skin, which was much tougher than I had anticipated; the needle bent slightly. I like to think that I'm pretty damn good at inserting needles into tough spots, but this was one of the most difficult veins I ever cannulated. For the umpteenth time in my life, I might have once again proved that no amount of skill can replace dumb luck.

I managed to get the needle into the vein, slid the catheter off, and started infusing IV fluids as fast as the narrow diameter allowed. I moved back to Tabibu's head and, with my left hand, held the mask I'd chosen as the likely best fit to her face, while with my right hand I injected a paralyzing drug into the IV to allow me to open her mouth and see the voice box that my breathing tube would pass through. I needed to breathe for Tabibu by mask until the drug took effect. Her jutting chin and wide nose made applying a breathing mask more challenging. I knew the drug was working when I could open her mouth without resistance. I placed the endotracheal tube through her mouth and into her trachea and secured it with tape to her lips. Turning on the ventilator, I watched Tabibu's chest rise and fall with every delivered breath, adjusting the settings until I was satisfied with the volume and number of breaths. The anesthesia gas was added cautiously.

Because she had not been eating or drinking, Tabibu's blood volume had dropped dangerously low, resulting in fluctuating blood sugar and oxygen levels and altering her electrolytes (the salts in the blood). In response to these changes, her breathing quickened and her heart pumped harder. We sought to reverse all these abnormalities by repeatedly measuring as many variables as possible and promptly making corrections until Tabibu was once again able to fend for herself.

With Tabibu motionless from my anesthesia, the surgeons gained the ability to expose a vein in her neck and
place a bigger-diameter catheter inside it. Now we could
easily sample her blood, monitor the results, and make
adjustments to return her blood volume, as well as the levels of the various salts contained in the blood, to within
normal limits. After every clinical adjustment, a new blood
sample was obtained and the surgical fellow and I reviewed
the results. While the numbers steadily improved throughout the day, Tabibu was by no means out of the woods. Her
condition could still change for the worse, and rapidly so.

When I checked on Tabibu late in the afternoon, her
numbers were continuing to improve, so I drove home.

THAT EVENING, SITTING WITH my family at dinner,
I announced to my eleven-year-old, animal-loving daughter Annie that there was something I needed to do and I
would enjoy her company. Annie dreamed of becoming an
animal nurse, and this was a once-in-a-lifetime opportunity. As we sped down the entrance ramp to the highway
pointed toward the city, Annie looked at me and guessed
we were driving to the hospital.

"Yes." I didn't elaborate, and she took a stab at the
reason.

"We're going to see a patient."

"Yes," I said.

"Who is it?"

"You'll see soon enough."

After parking the car, we walked up the hospital drive. But instead of turning right to the hospital lobby, we turned left toward the research center. Now Annie was very confused. She hadn't entered this area of the medical center before. In the hospital, displays oriented to children enliven the space. The research center lobby, by contrast, is a vestibule; the hallways are narrow and the walls are barren. The colors in the research center are neutral, and the only decorations are posters demonstrating research projects. The ceiling lighting fixtures, spaced wide apart, cast areas of brightness alternating with dark. The doors to the rooms are solid, without a window to give a glimpse of what's inside. This was not the welcoming place Annie had experienced in her previous visits to the hospital.

Annie followed as I opened the aluminum-cased glass doors that separate the floor into different departments. We climbed a narrow staircase up one floor. I opened the final door and stood back to let her enter first.

Annie stood motionless and wide-eyed—the same momentary paralysis that I had experienced earlier in the day. She broke free and ran to the side of the table. Tabibu was still not moving, except for a tiny twitch of her upper lip. I reached out and, with the outside of my extended index finger, rubbed the inside of Tabibu's exposed lip. She seemed to look at me, and she kept her lip raised. I believe she found comfort in that small gesture.

Annie asked if she could touch Tabibu. "Yes, you can."

Annie held Tabibu's head and, with her index finger, replaced mine and gently stroked Tabibu's lip.

My Annie is comforting a gorilla, I thought.

Blessed with a wonderful career, I reflected on how much more I might possibly experience. Could it get better? Every day holds the potential for a surprise, but this day delivered one. I crossed a divide between caring for humans and caring for animals. I'm not sure which was more special—watching Tabibu make even the slightest improvement or seeing the emotions on the faces of the people nearby. Most amazing to me, however, was watching Annie hold Tabibu and comfort this magnificent animal. It left me in awe. What joy.

THE NEXT MORNING, I WAS UP and out of the house extra early to allow myself time to evaluate Tabibu before beginning my scheduled cases. The results of Tabibu's blood draws indicated that we were reversing the dehydration, though the improvement was subtle. I met with the surgical fellow outside the door to Tabibu's room, and we agreed we were doing pretty damn well. But there was a long way to go.

The zoo vet approached us. "We'll have to make a decision soon about when to end care. The cost of keeping her alive will be prohibitive." Standing nearby was the zoo director, who, as I recall, remained silent.

The surgical fellow's response and mine came in stereo. The words flying from our mouths, the tone, and even the tempo were nearly identical—as if we had rehearsed our response: "I'm sorry, but Tabibu is under our care now. She's our patient, and we're making all decisions. Calling it a day is not an option."

"But we can't afford it."

"That's not your problem. It's ours."

I think I saw a relief-filled smile from the vet. Tabibu's welfare had become our responsibility, not hers.

There would be no charge for our professional services. Volunteers provided all other needed care, including nursing and respiratory therapy, and more than enough people offered their time to watch Tabibu twenty-four hours a day. The hospital administration, like the physicians, had chosen not to charge for the space or equipment for Tabibu's care, save for a tank of oxygen. They also chose not to publicize our unusual patient; understandably, I don't think they wanted to respond to the possible issue of giving free care to animals but not to needy and unfortunate humans, even though patients were never turned away for their inability to pay.

Annie returned with me to the hospital for the next two evenings and rubbed Tabibu's lip, which gave a little more response on each visit. Several days passed before the turnaround occurred. When Tabibu got well, it was like flipping a light switch. On Friday afternoon, four days after entering our care, Tabibu awoke. I removed the breathing tube from her mouth, and she sat up. Tabibu remained calm, demonstrating no agitation or aggressive behavior. I think she was still dazed from all that had happened to her and probably from the drugs she'd received. She was moved off the table and placed, uncaged, in an empty room-turned-pen. All the furnishings had been removed, and nothing hard or large enough to hurt if swung or flung remained. Blankets and pillows served for comfort, and she was able to move about with no chance of injury. Tabibu began to drink and to try some of her food.

The following morning, I received a call letting me know that Tabibu, growing increasingly active, had resumed her life in the great-ape house of the zoo.

A month later, the zoo graciously hosted a picnic for all those involved in Tabibu's care after closing one evening. My whole family accompanied me. It was a memorable evening. The polar-bear keepers stayed late to feed the animals, to our delight. Strolling the empty zoo was a treat. It seemed magical to wander the paths as night set in. The lights of the city provided an urban backdrop to the animals and exhibits, and the stars provided a show above.

I slipped behind the scenes, through the zoo's version of the automatic double doors, to the inner workings of the great-ape house. With only Tabibu's keeper escorting me, we walked behind the enclosure into the "Authorized Only" area that the keepers use for feeding and cleaning the gorillas. For one last time, I got close to Tabibu—this time with her sitting just beyond a pane of glass and preoccupied by something in her hands.

She was even more beautiful in her good health. There was no hint that she understood who I was or what role I'd played in her recovery. There was no hint that she recalled who had rubbed her lip for comfort while the ventilator breathed for her, or that she even noticed me. Sure, Tabibu's lack of recognition made me a bit sad. But anesthesiologists are commonly behind-the-scenes physicians, forgotten after providing care. And watching her in her home, healthy and seemingly happy, provided the cure for my sense of a lost friend. Every time I observe an animal separated by a pane of glass, memories of Tabibu are reborn.

CHAPTER 9

Errors Everlasting

SHE THOUGHT I SAVED HER DAUGHTER'S LIFE. I thought I screwed up. Years after caring for her child, I'm still unable to satisfactorily resolve this conundrum.

Toward the end of the day, deep in the OR suites, I was about to induce anesthesia to repair a broken arm when my pager vibrated. Pagers are both the bane of my existence and a treasure. I'm always reachable; that's the good. I'm always reachable; that's the bad. In medicine, I suppose there is no such thing as a good time to receive a page alerting me that somebody needs my attention. But there certainly are bad times. The callback number for this page was from our anesthesia prescreening office. I assumed that a nurse had a question concerning an upcoming case that she felt more comfortable having an anesthesiologist address.

I can't remember whether I was the on-call person assigned to field these types of questions that day, or if I got the call because the nurse just knew I was around. Or maybe my partner assigned to this task didn't respond. In any case, I was preoccupied with the patient in front of me, so I returned the pager to its home on the waistband of my scrubs and promptly forgot about it. Sometime later, maybe half an hour, I remembered the page and, feeling guilty, was compelled to answer in person. I found the nurse, apologized, and asked her what was up.

"There's an eleven-month-old scheduled for ear tubes and a hernia repair. The last time she was scheduled, she had pneumonia and was canceled."

"Healthy now?"

"According to Mom."

"Do we have a note from the pediatrician?"

"Yes. He thinks everything's fine."

"OK. Schedule her. Tell Mom I reserve the right to change my mind once we examine her. And I'll do the case myself." That's my standard response when presented with a case tagged with an issue. It's easier to provide the care myself and not cuff my colleagues with my decisions.

I gave the final answer and accepted the responsibility. The pediatrician was likely accurate—I wasn't second-guessing him—but sometimes the situation changes and the patient presenting for my care is not in the same condition as when last seen by the primary care physician. Kids are a host waiting for a virus, and the night is always ripe for attack. Evening ends fine and healthy. Morning carries a new illness and concern.

Several weeks later, my mind having erased all memory of that page, a now one-year-old girl named Jill arrived for ear tubes and a hernia repair. Jill's mom held her in her arms in bed space 9, and I didn't connect all the dots. I failed to immediately recognize that she was the previous pneumonia and phone call infant patient.

Then the case and Jill veered from the straight and narrow into one giant philosophical mud puddle.

MY ANESTHESIA RESIDENT PREPARED Jill for the case. When asked, he responded: "Her lungs are clear." I spoke with Jill's mom, who voiced no concerns. I attributed her nervousness as that of a typical parent surrendering a baby to an unknown entity, me.

As I reviewed Jill's medical record, I might have remembered the discussion with the screening nurse. A physical exam by the pediatrician followed her and stated that all was good to go. I questioned my resident regarding this infant's physical exam, and he said there were no issues. I talked with Mom, who had nothing to add. I believe I heard his mother say: "Take good care of her."

OR 11, a room reserved primarily for urology and ENT (ear, nose, and throat) cases, was the perfect room for this combined case. It's not just small; it's tiny. For this combination of procedures, besides the microscope for the ear tubes, the equipment was basic. The room fit. With Jill sitting on the OR table, my resident stood at Jill's head in the anesthesia nest while I stood at Jill's left arm, applying the monitors as I had many thousands of times before. I placed the puppy dog stickies, the gummy EKG pads decorated with mini dog cartoons, on Jill's chest, then wrapped the glow Band-Aid (the pulse oximeter) around her left thumb. My resident placed the mask over Jill's face and began to induce anesthesia by gas.

Jill remained sitting up while her anesthesia was induced; then we laid her down. I moved into the nest while my resident went around the OR table to start an IV. With the IV in place, I removed the mask from Jill's face momentarily to allow the resident to return to his place at Jill's head. The pulse oximeter reading dropped—not dangerously, but any drop was not right. The level of oxygen in Jill's blood was lower than it should be. The mask was off only a few seconds, and this drop came awfully quick. Too quick. I noted it as the case moved to its next stage.

Normally, placing ear tubes is done with not much more anesthesia intervention than holding the mask to the patient's face and allowing the patient to breathe without assistance. But since there was also a hernia repair to be done, I decided to use a plastic airway device, an LMA (laryngeal mask airway) slipped into the back of Jill's mouth. In a case such as this, the LMA eliminates the need for my hand to hold the face mask. When the mask was removed from Jill to place the LMA, the tone on the beeps of the pulse oximeter dropped again, indicating lower-than-anticipated oxygen levels. This was real, not an artifact.

Of the monitors I use, the pulse oximeter is the attention-grabbing commander. The glow Band-Aid is a single-use light and receptor that wraps around a fingertip and is held in place with an adhesive elastic strip. The reusable version for adults is a larger finger clip. The patient's finger glows from a painless red light source that shoots through the skin and tissue and is received by a photoreceptor on the opposite side. The red light is actually two separate lights, one red and the other infrared.

Hemoglobin, the oxygen-carrying component of blood, absorbs the light differently, depending on whether it is loaded with oxygen or not. The hemoglobin carrying oxygen absorbs the infrared wavelength, while the naked hemoglobin absorbs the red light. The monitor provides a number based on the percentage of hemoglobin that is oxygen rich. Healthy people have pulse oximeter values in the midnineties or higher. One hundred percent—hemoglobin that is fully loaded with oxygen—comforts me the most. The monitor also displays a signal strength tracing and sounds a tone with a pitch that decreases as the oxygen saturation decreases.

The tone is the attention grabber. Even a single percentage point drop in saturation results in a perceptible change in pitch that twists the necks and turns the heads of everybody in hearing range toward the monitor.

The peaks and troughs on the monitor don't necessarily correlate to the patient's condition; many external variables contribute. The readings on the monitor can even be deceiving. Far too often, when the readings waver, all eyes home in on the monitor screen. I've resorted to placing a towel over the screen to stop residents from watching it instead of the patient. Sometimes I sermonize. "What do you call a person who stares at the monitor all day? A statistician. What do you call a person who stares at the patient? A clinician. Which would you rather be? You don't need to go to medical school to stare at a monitor."

It is true that the pulse oximeter sees much better

than the human eye. I've observed skin color in patients under anesthesia for decades. When I test myself against the pulse oximeter, I'm unable to determine a dropping oxygen saturation until the pulse oximeter reads eighty-seven percent. But it would be wrong to think that this monitor drastically improved health care outcomes. To date, the pulse oximeter has not been proved to be effective in bettering outcome.

Still, the pulse oximeter provided the impetus for me to identify Jill's problem and to make a diagnosis that was potentially lifesaving.

Hearing the tone of the beeps on the pulse oximeter drop, I suddenly realized that Jill was the infant with a previous pneumonia that I had been consulted about. I was left only to ask myself how I'd been so damn oblivious.

HEALTHY INFANTS DON'T JUST contract pneumonia. I can't count the number of times I've taught residents-in-training that wheezing and pneumonia in infants are signs of a greater underlying problem. With my superspecialization of anesthesia for congenital cardiac disease, heart defects are the most common cause I see. Jill might have an undiagnosed heart defect. Failing to pursue the pneumonia issue at the initial page was my first mistake. I had blindly trusted the pediatrician and his assessment. Not listening to Jill's lungs myself before the OR was my second error. Trust but verify.

I reached for my stethoscope, knowing that my exam would most likely reveal abnormal lung sounds. I anticipated hearing the little popping sounds that indicate air is entering blocked lung alveoli, the little terminal sacs that exchange oxygen in the air with the blood. This is the sound made when the lungs have too much blood flow from a maldeveloped heart and the sacs are filled with extra fluid. The sound, termed "rales," is made from the air sacs popping open with the rush of air. Maybe I would hear wheezing, the sound that air leaving the lungs makes when the airways become narrowed by muscle in the walls tightening. Wheezing would imply asthma.

The ENT surgeon finished placing the ear tubes and moved off to the side to allow me access to tiny Jill's chest. Her breath sounds weren't as I expected. I didn't hear abnormal breath sounds; I heard *no* breath sounds throughout the left lung field. Something was blocking air from entering her left lung or was deadening the sounds from being transmitted through the chest wall into my stethoscope. I no longer suspected undiagnosed heart disease; that would affect both lungs. This was all about Jill's left lung.

I told the staff in the room that I wanted an X-ray in recovery. The surgeons, both the ENT and urology attendings, looked at me.

"We've got something here," I announced. The rest of the room just stared. Normally, the oxygen content of the gases I provide the patient is twice that of room air. In the oxygen-enriched gas that Jill breathed, her pulse oximeter reading throughout the procedure remained normal, her hemoglobin saturated. Now, in the middle of the surgery, canceling the procedure already under way seemed counterproductive. The only choice was to move ahead.

Jill's hernia wasn't repaired earlier because the pediatrician had heard some wheezing during the presurgical physical exam. The goal was to optimize Jill's health. The case was postponed.

The remainder of the case progressed well, but one more test at the end of the hernia repair surgery reconfirmed what I already knew: Jill's lungs were not filling her blood with the correct amount of oxygen.

In recovery, the chest X-ray I ordered was taken. Before the results were ready, Jill's mother came to be with her in recovery. I spoke with her.

"I ordered a chest X-ray. It wasn't anything dangerous, but she dropped her oxygen levels quicker than I anticipated. It might be related to her pneumonia, and I just want to make sure."

"Thank you!" her mother gushed. "Finally someone is listening to me. I've been telling everybody that something's not right. They won't do anything."

"Well, I'm listening. We're getting the X-ray. We'll figure it out."

And so the story continued with an almost expected plot. I wasn't surprised at Jill's mother's claim that no one was listening. The vast majority of a pediatrician's office practice is well-child checkups, runny noses, colds, and vaccinations. And I wasn't surprised by her insight that something was not right. Mothers always know best. I was surprised by how blindly I had acted up to this point. When moms say something's wrong, it's true until proven otherwise. This is a belief I was taught early on. I was surprised I had missed so much. All the signs and symptoms were there, and I hadn't completed the picture, the story line. My curiosity had finally been piqued in the OR, but to me, this was too late.

When the X-ray was posted, I stared at it wide-eyed. Something was definitely wrong; it just wasn't what I guessed it would be. There was no pneumonia. Jill's chest looked as if a basketball was resting right in the middle of the anterior view. It wasn't a misshapen heart that would occupy the X-ray area in question. The radiologist's interpretation was telling. The chest X-ray revealed a "prominence."

A more definitive X-ray followed and coincided with a flurry of activity that found Jill first admitted to the hospital (hers was supposed to be an outpatient procedure), then soon after transferred to intensive care. A blood test confirmed that Jill was very, very ill. Jill was suffering from cancer.

That underlying problem revealed itself with even greater clarity when Jill's belly swelled with a blood clot late in the evening. Jill had too few platelets, those bits in the blood that are the first line to forming a blood clot, and the cuts of the hernia repair were just large enough to push her over the bleeding edge.

I lost track of Jill after she left the hospital. That is, I knew she was in treatment and responding, but I had no more contact.

A year and a half later, as I stood speaking with a colleague at the entrance to our procedure suites, I noticed an older woman standing across the hall, staring at me for an uncomfortably long time. She turned away and a moment later returned, following a younger woman who was pushing a familiar stroller. Before I could register who it was, I heard the older of the two say to her daughter: "That's the doctor who saved Jill's life." While most would be proud of a praise sung like that, I remain unconvinced. What mattered at that moment, though, was that Jill had completed her therapy for cancer. She was cured, and she looked marvelous.

If I had connected all the dots, put all the hints into proper perspective, might Jill have died? That was, and continues to be, my dilemma. By trusting without verifying, I placed Jill into the OR before making a diagnosis, albeit a crude one: no left-sided breath sounds. If I had verified the resident's exam by listening myself to her breath sounds in the pre-anesthesia area, I would have recommended the case be canceled, and Jill would have been sent back to her pediatrician, who had also failed to connect the dots, or to listen to this forlorn mother.

Missing the diagnosis is not a major blunder on the pediatrician's part. Statistically, this is likely to be the only such case that he will ever have. Jill might not have received care soon enough. It just happened to be that, by not performing at my most diligent, I saved her life. That reality has made me uncomfortable ever since. My story remains without an end, and my dilemma unresolved. Returning to the skill-versus-dumb-luck adage, does a superior outcome justify a less-than-stellar means?

As a side note, that pulse oximeter reading proved to be the vital sign leading to the discovery of Jill's illness. Under my care, Jill never turned blue. The oxygen level in her blood never dropped low enough for her to turn from pink to cyanotic. By observation alone, Jill's color didn't change. But the monitor tone did, triggering my suspicion and search for a cause.

Having thought about it, I realize that to claim I saved a life is too bold. I didn't. I expedited a cure.

A LEGEND IN MY CAREER, Frank Seleny, was the anesthetist in chief who accepted me into his program as a pediatric anesthesia fellow, trained me, and finally hired me. One day early in my career, he pulled me aside and advised me that it would take ten thousand anesthesia cases before I would understand the limits of my ability. To paraphrase Donald Rumsfeld: There are known knowns, and known unknowns, and then there are unknown unknowns. Frank taught me the value of recognizing and limiting the unknown unknowns.

The next step toward expertise is understanding that shortcomings should not provide avenues for criticism, but motivation for change. This revelation struck me midcareer. It called for me to open my mind to become willing to seek continuous improvement. To recognize that mistakes are not always a sign of incompetence that should be buried (no medical pun intended). As Niels Bohr, the Nobel Prize–winning physicist, put it: "An expert is a man who has made all the mistakes which can be made in a very narrow field." I believe I've made them all, plus one.

One mistake from early in my career remains at the

forefront of my memory—not for the injury to the patient, but for the damage to my ego. I was working with the Learned One, the brightest attending and the most particular about process. The surgeon was another legend of my career, Casey Firlit. At the end of the procedure, the patient took an unexpected giant breath just as I switched off the ventilator that had delivered his breaths throughout the case. There was but a fraction of a second for this to occur, but it was like trying to take a breath with a plastic bag tied tightly around the head. His chest sucked in as he used all the energy he could muster to take a breath. He developed a strong negative pressure inside his chest that injured his lungs, causing pulmonary edema. He emerged from the anesthesia, I pulled the endotracheal tube unaware of the injury brewing, and pink, frothy fluid poured from his mouth.

My heart sank as I realized I was the cause of his pulmonary edema; fluid leaked out of his lungs. His oxygen level started dropping, and the breathing tube needed to be replaced to create positive pressure in the lungs and stop the fluid leak. The Learned One just looked at me, but the pain was piercing. My care, or lack of it, had caused a patient to be injured. Casey said little. The patient was transferred to intensive care, and I had my first very difficult discussion with a family. Several hours later my patient was awake and fine. But I had let three of the most important people in my life at that time down: the patient, the Learned One, and Casey.

THE PRIDE I FEEL after a case of a critically ill patient ends successfully fills me to bursting. It's true that not all of these cases end well. That's the cost and the burden I carry as those patients forever inhabit my memory.

Two times I have been accused of medical negligence, and both anger me. In both cases I was part of a team that tried to preserve a life, and both times I was soon dropped from the suit. But the accusation is enough to hurt, and regardless of the lack of proof, I still must list both cases when I apply for privileges anywhere.

Far worse and more haunting than the two claims of negligence was the case of Spencer. Multiple congenital defects required multiple surgical procedures on multiple parts of his anatomy, from the top of his head to his hips. Early on, a tube was inserted into the trachea in his neck, bypassing a blockage above, and it remained until a surgically improved airway was obtained.

He came for a finishing procedure, the tracheostomy removed and healed. I placed the breathing tube with difficulty and an improvised technique. The procedure went as planned, and I transferred Spencer to the intensive care unit with the breathing tube in place. My intention was to allow Spencer to recover fully from any and all residual medications that might alter his breathing, and I left with orders to contact me prior to removing the tube.

The following morning, as I was in the procedure suite providing care, the endotracheal tube was removed without my knowledge. Spencer died. While I stood one hundred feet away, Spencer died. Nobody could manage his airway in the manner I did. I failed to be clear or persistent enough to prevent an overzealous physician from going rogue. Personal beliefs prevented the family from pursuing a claim.

SOMETIMES THE PAIN of failure extends past the patient.

The fact that anesthesia is iatrogenic—I am not a healer—elevates complications to another, higher level of guilt. One of my worst complications came through an extension of my hands.

A misshapen head is a curse borne by the owner but noticed by everyone. Surgeons have gone to great lengths to correct a deformed skull. Not all have achieved the desired effect. Long before I came to meet Carter, he underwent surgery and therapy that left him with defects—holes—that exposed his underlying brain to trauma. A strike in the right place would not be deflected by the bony skull. With his skull exposed for a lengthy procedure, and with significant blood loss, Carter's temperature dropped. My resident became concerned, despite my assurance not to worry—that near the end of the procedure we would correct the hypothermia. Without discussing his plan with me, the resident placed a warm compress on Carter's skin. When the surgical drapes were removed, a patch of skin was left burned and leathery.

I wanted to let go a primal scream. My resident's action had been stupid. Even though he was the one to place the compress, I was the one accepting all responsibility and the one to accept the blame. I didn't yell. There were no "goddamnits." There was a parade of "shit, shit, shit, shit, shit." My talk with the family gave new meaning to the walk of shame.

Carter's family was, beyond all expectations, reserved and understanding. He and his family refused to file a malpractice claim. Instead, when he returned for some follow-up surgery on his head, the plastic surgeon repaired the burns. I cared for Carter several more times at the family's request. After the third or fourth procedure, I asked his family: "Why do you always request me? This was one of my worst complications."

The response at first stunned me, then became clear: "That's easy. Now, every time he goes to the OR, I know he has a guardian angel."

They were so right.

CHAPTER 10

In Wait

"GODSPEED."

Despite having plenty of time to find the right words to express my love, and despite my abundant experience in these situations, my mind froze. In that moment, at the point of separation, all I could muster was that single word. Do better words exist?

I surrendered my son. *They took him from me.* I remember that exact thought racing through my mind as we were separated. Whether one's child is three years old or thirty (as my son was) or older still, the anxiety a parent feels when the child undergoes a medical procedure does not lessen. With a kiss on his forehead, a one-arm shoulder-to-shoulder bump hug, and that simple, single-word prayer, he was wheeled away.

I considered telling Jason "Good luck" but caught myself and stopped. I considered how I move my patients to the operating room. To families who say "Good luck" at the moment of separation, in my role as the anesthesiologist I always respond: "Luck is for sports and betting. Mine is a world of skill." During surgery, luck is not a trait to depend on.

The back of the cart where Jason lay was raised slightly, and as he was wheeled down a hallway I could see his shoulders extending over the sides and the short brown hair on the top of his head swaying slightly to the left, then back. I imagined he was speaking with his attendant. Jason's larger-than-life personality grabbed the attention of everyone near him. I prayed that on this day this trait would not be taken from him. His form on the cart shrank with every step until he drew near the automatic double doors, which snapped open and then shut as he entered the operating room.

I was caught on the unfamiliar side of an all-too-common scene, one that repeats itself across the country 150,000 times on an average day. Spouses, mates, parents, children, and friends exchange hugs and kisses with a loved one who is about to enter into an unknown, sequestered space—where senses are reversibly altered to allow a body to undergo an invasive and painful procedure out of medical necessity.

———

OVER MANY YEARS, I trained to be the physician accepting the responsibility of providing safe care to patients while ensuring their comfort during an invasive medical procedure. Safety always trumps comfort. Specializing in anesthesiology and pain relief, I normally wait on the inside, preparing to care for those passing through the double doors pasted with colored warnings, black on yellow and red on white: "Authorized Personnel Only." Care may last from as little as fifteen minutes to more than fifteen hours, but the length of time doesn't mean much. The shortest cases may be the most critical, while the longest may be benign. Regardless, my pledge as an anesthesiologist—in concert with the physician performing the procedure—is always to return patients to their families in better condition than on separation.

On that day, with my son, I was on the other side. His dream of being a physician lay at risk. A clump of abnormal blood vessels of underwhelming size on imaging—but potentially life-altering, if not deadly— sat in a precarious location near the part of the brain controlling movement, poised to burst and possibly deprive him of the use of a hand and his career. Jason chose bravely to have the vessels removed. He needed to know—not to live wondering every day if it would be his last. He chose surgery.

In a preprocedure room no different from the thousands I had entered before, Jason rested on a cart positioned in the center with just enough space for one person to move along each side. Three thin walls enclosed the room, with the fourth consisting of sliding glass doors and privacy curtains. The lighting was sterile fluorescent, blue and cold, lacking comfort. Three people, my son included, crowded into the prep room, and I was not one of them. I peered in through the doorway as I stood outside the circle of trust, while overcome with a complete sense of helplessness.

Excluded from the discussion of intended care—not checking his chart, test results, and consents—I remained outside the glass doors and in the hall, dodging the flurry of movement around me. Receptionists escorted patients and families to assigned spots, nurses readied patients, checklists, and all, and physicians performed physical exams, obtaining consents and entering notes into the record for what was to follow. In his room, Jason appeared unaffected by the commotion. Perhaps he was calm because his decision had been made. His responsibility was complete. Now came the time for the anesthesiologist and surgeon to step up.

In the halls of the pre-anesthesia area, people are always nearby and talk is rampant—especially in the early morning, with all the first-scheduled cases vying to start. Privacy is nonexistent, curtain or not; everyone hears everything. Not within the circle of care, not a member of his health care team, and forced to stand outside with all the others denied entry, I watched my son be led away by someone I really didn't know.

I NEEDED ANESTHESIA MYSELF ONCE.

It was our annual Turkey Bowl, a friendly Thanksgiving morning noncontact football game with friends and colleagues. A rush of testosterone, a crash with the largest opponent and then with the frozen ground, and a midcareer anesthesiologist became one of the forty million who receive anesthesia every year. I expected this reversal of role, doctor-turned-patient, to fuel a revelation and ignite great changes in practice. It didn't.

I stood in amazement as my colleagues, including surgeons, radiologists, and anesthesiologists, wanted to put me on one side of a fence and yank my arm back into place from the other side. They had watched somebody repair a dislocated shoulder this way on some TV show.

"Are you nuts?"

Their response was a shrug.

I found myself in a pre-anesthesia space as a patient. I not only knew my anesthesiologist, I had trained him. I had already placed my faith in him. My three minutes of face-to-face time expanded to five, but not with the exchange of useful information. We just shot the bull.

As a believer in the philosophy of don't do to others what you won't do to yourself, at least to the extent possible, I have stuck myself for an IV. Slick and quick is my motto. I dedicate myself to declaring "I'm done" before the patient flinches. The discomfort can be lessened, though, by a technique that mimics the whoosh sound produced when a soda can is opened. The J-Tip is a tiny syringe filled with a small amount of local anesthesia in saline. A carbon dioxide propellant creates the sound of a soda can being opened. Without a needle but with the push of a lever, the mixture is painlessly propelled by the gas into the skin. I demonstrate this technique on myself to reassure dubious patients. The resulting pencil eraser–sized welt is deadened to pain. (My issue with this technique is that it encourages people who start IVs to abandon the slick-and-quick approach and be less efficient.) Then the IV stick itself doesn't hurt.

With the IV in, I watched as morphine flowed into the line and thought: *That's a hefty dose.* With so much access to so many drugs, anesthesiology leads the specialties in substance-use disorder. People who have been hooked on these drugs say that if you see a bullet heading straight for your eyes, grab for a narcotic; there's no more comforting feeling. So I was expecting a rush. But I felt nothing.

The surgeon said: "You can't go home today. There'll be too much pain."

I didn't argue. I expected substantial discomfort. At issue was the use of patient-controlled analgesia, or PCA, to control my pain. Policy entered decision making. Substandard care becomes possible when policy dictates physicians' orders for care. Because my procedure was classified as outpatient surgery, I could stay in the hospital no more than twenty-three hours. Extending the stay would cost me thousands of dollars via an insurance denial as, by definition, I would transition from outpatient to inpatient care. The effect of PCA might last too long and potentially delay discharge. Four hours after surgery, I pressed the nurse call button for pain relief and was offered a pill. I expected pain relief via as-needed intravenous medication. Hell, I could be home medicating myself with better effect than that single pill offered.

The orthopedics resident stuck his head in the doorway to my room and, not wanting to debate the issue, asked: "Doctor Jay, what do you want?"

"Four mg morphine IV q 2 hours and add 30 mg Toradol IV now, please."

The blunder was mine. I failed to follow my own recommendation, to discuss postprocedure pain relief before the procedure.

I received the morphine and Toradol and, as the sun

rose the next morning, my surgeon walked into my room with an announcement.

"OK. OK. I don't want to hear about it. You could have gone home last night, and I don't want to hear anymore."

I have no idea what the ortho resident or nurse told him. I thought I'd been polite.

Before entering surgery, I knew my options for postoperative pain relief. The first choice, regional anesthesia (local anesthetics injected at the base of the neck bathing the nerves to the arm) would have worked. But I had miscalculated. The pain wasn't horrible; it just could have been treated better. The second choice, narcotics, offered immediate pain relief. The accompanying sedation, a side effect not always desired, was something I wanted to help me sleep. Shoulder injuries make it hard to be comfortable in any position. By IV, the narcotics would take effect quickly, and I could transition to meds by mouth when comfort was achieved.

THE ROLE REVERSAL I experienced with my son was worse than the one I experienced personally, because it wasn't my body or my life. It belonged to a loved one undergoing a procedure.

It's nerve-racking to surrender control of someone so

close—worse than surrendering myself to the knife. My inability to control the situation or even assist frustrated me, and my inside knowledge only made things worse. Though I appeared outwardly calm, a fire raged inside me.

Patients choose their internist and surgeon, but rarely do they choose their anesthesiologist. Jason hadn't selected his anesthesiologist. Neither had I. Although I made no request, I knew his anesthesiologist as a colleague, and I knew that he had been intentionally selected for his special expertise, neurosurgical anesthesia.

Being left on the outside as the OR doors closed, with Jason more than out of reach and sight, left me cold. My soul split as my father-half whispered that it was time to sit and wait and pray, while my physician-half could make no sense of leaving my son.

I routinely advise families not to stay in the waiting room during a loved one's procedure. I'll find them if I need them, I tell them. Let me do the worrying. I didn't follow my own advice and was overcome with waiting-room paralysis. My body felt welded to the chair, with my feet twitching, in constant tapping motion. I didn't heed my own philosophy and surrendered to the thought that if I moved, even to stretch, in the microsecond that I was not sitting in my chair I would miss attempts from the OR to reach me.

I rose once for coffee, but only with the assurance that

my wife, his mother, remained in her place ready to receive any news. The coffee machine was in plain view. While filling my cup, I maintained visual contact with my wife and with the control desk receiving and distributing information. Still, the thought consumed me that the receptionist who had marked on a map of the waiting room the chair I occupied would look and see it empty. There was no bathroom break. There was no need for one. I think my kidneys shut down in sympathy.

I observed the other people in the waiting room. Some sat with prayer books in hand, their anxiety bared for all to see, while others sat with novels or tablets, pretending to read. Some sat in groups with food spread out in preparation for a marathon wait—a morbid picnic. I sat down in a corner chair, leaning forward, staring at my hands clasped together between my knees.

I knew the actions taking place behind the double doors. Sitting in the waiting room, I envisioned Jason's operating room. I calculated how long he'd been gone and predicted what was taking place. I saw myself, or maybe willed myself, to be standing in my son's anesthesiologist's place. I wanted to be the one making the decisions. I wanted to be the one pushing drugs into Jason's IVs. I wanted to control his vital signs. I wanted to observe the blood vessels exposed by the surgeon bouncing with each heartbeat—an indication of the vitality of my son's heart, blood pressure, and blood flow.

I saw myself noting the color of the blood in the surgical field. Bright and red. Filled with oxygen. That's good. I saw myself turning to look at the stacks of waveforms of the colored tracings crossing my monitor and the numbers to the side representing all that I had just assessed in the surgical field. I was watching the surgeon's hands, predicting his next move, observing the tips of the instruments, ready to act, knowing that as the tips disappeared, unseen structures might be breached or severed and complications might occur. I was ready to counteract anything that might go awry.

A mind in the waiting room wades uncontrolled into all avenues of thought. I found myself counting backwards through my life. I reached all the way back to my dream of becoming a doctor. Raised in a blue-collar environment, I saw nothing more of medicine as a child than begging my doctor not to give me a shot. As a prospective medical student, I asked this same man to perform my pre–medical school physical exam for a discount, as I had no money. And I had no idea what lay ahead.

I have witnessed feats of prowess by all in the procedure suite, and I have experienced a miracle or two. But on this day, all of my knowledge and experience didn't help. I yearned to sail through those doors and tend to my son. Like everybody else waiting, I wanted my son back.

AFTER SEVEN HOURS, Jason's surgeon sauntered into the waiting room—a good sign. The surgery was a success. "We're done," he said, "but it'll take another hour to close." What he really meant was that he had completed his part, the intricate part, the resection of the vascular abnormality. Now it was left to the junior surgeon, fellow, and residents to close the hole in my son's skull and the wound in his scalp. I breathed a huge sigh of relief and thanked the surgeon.

It was much longer before I could reunite with my son. Bureaucracy delayed Jason's transfer from the OR; with the sun low in the sky, his room was not yet ready. To lessen my anxiety, my anesthesiologist colleague texted me a photo of Jason wide awake, giving thumbs up, in the OR.

He returned to me.

Paper Cranes

PAIN IS A MEDICAL ORPHAN. PERHAPS BECAUSE it has traditionally been considered the consequence of disease or injury, not an illness in itself, and not specific to a body organ or site, no single specialty has accepted, as a pressing goal or major responsibility, a commitment to the elimination of pain. Perhaps there's a little too much "man up" sentiment out there, embracing the words of Nietzsche: "That which does not kill us makes us stronger."

The roots of pain relief evolved from the search for pleasure via altered consciousness, as the Sumerians discovered from the "joy plant" over five thousand years ago. Opium from the seedpod of this poppy plant, the parent of the contemporary narcotics morphine and heroin, was first isolated and used in the Fertile Crescent between the Tigris and Euphrates Rivers, in the area known then as Sumer (present-day Iraq). Recognizing opium as more than simply a pleasurable escape, the ancient Egyptians used it for the distinct purpose of pain relief. Little attention was given to discovering new pain relievers until the late 1600s, when opium added to alcohol (laudanum) was introduced in the Western world.

The search for pain relief accelerated in the 1800s with the formulation of morphine, the name derived from "Morpheus," the Greek god of dreams—a reference to its sleep-inducing properties. Morphine and aspirin (which came later and is derived from the bark of the willow) were used with the specific intent of reducing pain. Subsequent research has led us to understand that the opiates—opium, morphine, heroin, and the newer synthetic narcotics—act on the mu and kappa receptors in the spinal cord and brain to lessen pain. Unfortunately, they also interact with dopamine in the brain, creating pleasure, thus causing them to be dangerously addictive.

The discovery of anesthesia in the 1840s changed the landscape of medicine, allowing invasions so horrific that pain relief after emergence became a requirement. Still, it wasn't until after World War II that anesthesiology laid claim to the responsibility of pain relief under all circumstances, and introduced the concept of pain clinics, which are still expanding in size and scope.

A LARGE PORTION OF MY anesthesia tool chest is filled with pain relief drugs derived from plants. The poppy, the coca plant, birch bark, cannabis (as an anti-inflammatory), and coffee (for me, good effects on multiple organs and the downside not obvious). All these drugs and plants are steeped in a rich history that enlightens and invigorates me, and when used creatively, they can ease or eliminate physical pain. The pain I encounter, however, is not always limited to the body.

Freedom from pain should be an undeniable human right in all places, under all conditions, and at all times. I didn't accept this principle *in toto* on entering medicine or anesthesia. I reached it in a far-too-long maturation process.

Throughout my early years of practice, I accepted the prevailing attitude of anesthesiology as a specialty: out of sight, out of mind. The philosophy in my training was that patients discharged from anesthesia care were of no further concern to the anesthesiologist. The responsibility for pain relief fell on the service and physician whose procedure had caused the discomfort.

My newfound principle grew through a series of experiences that sometimes we in medicine overlook as we become enamored with the need to use the most technical advances to treat the few, while rather simple methods are readily available and underutilized to care for the many.

After I'd been an attending anesthesiologist for several years, supposedly already knowing, mature, and experienced, I was invited to China to share my expertise about pediatric anesthesia with fellow anesthesiologists in Shanghai and Beijing. Thousands of miles from home, I walked through the halls of my host hospital in China, my head bobbing with every passing room, my eyes overcome with curiosity and irresistibly drawn into each.

It was a surprisingly modern hospital—at least, given my preconceptions of China. The hall was long, with many rooms on both sides, and the walls were painted hospital white. One room, in particular, stopped me. Dozens of paper cranes—perhaps a hundred, maybe more—hung from the ceiling and floated over the bed. Under these cranes lay a listless teen, his hair sparse, with a smattering of sores about his lips. This boy most likely suffered from leukemia. Cancer, and more specifically its treatment, has a way of stealing identity and gender. He returned my gaze, his only movement a slight tilt of his head. His mother, short and stout, stood on the far side of the bed staring at me, a foreign invader, with a look of part anguish and part suspicion.

I was to lecture on the principles and practice of Western anesthesiology. I had been under the impression that the Chinese health care system had imploded as a result of the country's cultural revolution and had fallen decades behind current medical practices, and now it was attempting to catch up to current worldwide standards. But standing in that hallway and looking at those cranes casting shadows on a pained teen opened my mind in a different direction.

In this modern, state-of-the-art facility, the boy likely received a treatment not much different from what my hospital would have provided: a lytic cocktail of chemotherapeutic medications intended to annihilate the cancer cells. The coincidental loss of normally rapidly replenishing cells included the hair follicles, causing baldness, as well as skin cells around the mouth, resulting in painful oral ulcers. Lacking a head of hair, this teen had a barely identifiable gender. But his pain was clear. The crafted cranes represented an amalgamation of medicine and culture, and symbolized the hopes and prayers for the teen's future. They had likely been folded with patience by a family desperately imploring the beliefs of a culture thousands of years old to intervene in a manner that medicines could not, to save the youth's life. Each folded crane added to the chance their wish would come true.

Those cranes taught me that I was not so much a

teacher, but a student learning about the power and significance of Eastern beliefs. They taught me how much I didn't know about Chinese culture and medicine. I also learned that the origami art form that I associated with the Japanese actually began in China thousands of years before medicine stepped in to treat cancer. That paper was a sign of wealth, and the carefully folded cranes beckoned good luck or recovery from illness. The color of each crane was specific for the wish at hand.

I found expanding my cultural knowledge fulfilling, but my first truly seminal moment came on a different ward. I approached a room, outdated by my standards, with four beds. Only one was occupied. In the farthest corner from the door, the farthest possible distance from the nurses sitting at a nursing station outside the room, a young boy writhed in a bed, his arms and legs tethered to its four corners to prevent him from hurting himself. A day or maybe two earlier, he had undergone surgery to repair a chest wall deformity. Rolling in bed, he clearly wasn't attempting to injure himself, but rather was experiencing excruciating pain. Unable to comprehend why his extremities were tied down, he struggled and pleaded for comfort and peace.

To me, the problem was perfectly clear: this boy needed more narcotic pain relief medication. However, my attempts to explain the importance of more com-

plete and more compassionate pain relief, and to argue
that healing would improve with strong analgesia, fell
on deaf ears. The Chinese physicians I met accepted
paper cranes and acupuncture (I attended a clinic about
attempting to repair nerve injury by acupuncture), and
they performed invasive surgical procedures. But they
seemed impervious to my urgings on pain relief.

WHEN I RETURNED HOME, one of the sponsors of my
trip hosted a dinner. Physicians from a variety of specialties attended, and after the obligatory cocktails and
conversation came the announcement that the organization had raised sufficient funds to cover the surgical
repair of a heart defect in one Chinese child. I asked
about the heart defect and recognized it was a type associated with Down syndrome. I summoned the strength
to stand up and press for the funds to be diverted from
treating a single patient to relieving pain in many. I
knew this was a risk. If I presented my views cogently, I
would win the opportunity to provide my expertise on
pain relief to the benefit of society. But if not, I would
lose the chance to interact with this group again.

Mine was not an issue of ethics or of providing care
to people suffering congenital abnormalities. Thousands of my patients have suffered from genetic or congenital defects. I knew from experience that a Chinese

child with such a defect would, in all likelihood, do well with the surgery and correction, but would return to a culture that refuses these children societal acceptance and places them in an orphanage. During my trip I did not see one person suffering from an obvious genetic syndrome, Down syndrome included, outside of a hospital or orphanage. Not in a market, shopping mall, or restaurant, or even on the street.

The China trip forced me to reassess my efforts at alleviating pain. I wasn't anti-child—especially not anti–Down syndrome. I was pro–pain relief. My rediscovered passion concerned the ability to provide successful care, and to allocate health care dollars and effort for the greatest benefit. With little concern regarding expense or compensation, the United States provides care to anyone and everyone. After that haunting image of the boy in Beijing, painfully writhing, it became my mission to urge these benefactors to allocate their limited resources toward the benefit of many over the care for one. The boy begging for relief from pain was emblematic of an overarching tendency in China to undertreat patients. Although caring for a single afflicted child allowed photo ops and plenty of favorable press for all involved in the funding and care, redirecting the funds to educate many Chinese physicians on the use of inexpensive pain medications could alleviate widespread suffering in the long run. (The cost of treating significant pain today averages $1.67 per dose.) I urged treating thousands over treating one.

I lost.

Still, I remained resolved to stand by my commitment to alleviate pain, or at least to attempt to relieve all pain, beyond the walls of the procedure suite, the perceived boundary for many anesthesiologists.

Relieving pain would seem to be a primary goal for the anesthesiologist. And analgesia isn't difficult to understand—"Do you hurt?"—or to treat. But for a long time, and like many anesthesiologists, I simply put patients out of mind once they were out of sight.

Witnessing the Chinese boy's suffering was the final solidifying moment in my understanding of the importance of pain relief before, during, and after medical procedures. Initially, my concept of pain relief was limited by preconceptions and prejudices that kept my mind closed. Pain itself remained an enigma to me. The thesaurus lists over sixty words interchangeable with "pain." That doesn't include all the qualifiers—"dull," "lancing," "constant," "throbbing," and many, many more. The challenge of defining pain compounds the challenge of eliminating it.

One tool I use in teaching—not so much for its educational or even patient care value, but to grab the attention of those I'm speaking with—is an outdated instrument for assessing pain called the dolorimeter (*dolor* is Latin for "pain" or "grief"). Described in an issue of *Time* in 1945, the dolorimeter was a goofy-looking, double-pressure, pain-inducing contraption appropriately lost in history. Unlike the MAC (minimum alveolar concentration) of the volatile anesthetics, which can be succinctly measured, pain is an elusive property. All current pain measurement instruments continue to rely on a subjective assessment by patient or health care provider—a guesstimate at best.

The current pain scales are either visual or numeric. The two scales known as VAS (Visual Analog Scale) and FACES consist of cartoon drawings representing circular faces that range from happy (a smiling face) to sad (downturned lips) that are intended to assess pain in young or nonverbal patients. The VAS dates back nearly a century, with most current research efforts claiming a 1923 paper as the origin. On a line with "No pain" at the start and "Excruciating pain" at the end, a mark is put where the pain is ranked. The numeric scales rank pain on a scale of 0 (no pain) to 10 (the most excruciating pain ever).

In contrast to these simple forms of measurement, Donna Wong, by all standards a nurse extraordinaire, and Connie Baker, a child life specialist, teamed up to produce and publish their FACES scale in 1988. Then the debate begins: Is a 6 a 6 for everybody? At which ranking and with which drug and at what dose should a patient be treated with pain relievers? My intention is for every patient to score pain a zero, but that goal is unrealistic. Without additional attempts at pain relief, nobody leaves the recovery area with a score greater than the second face on the scale or a numeric of 3/10.

This little boy halfway around the world in pain and tethered to a bed awakened me and spurred my growth into an advocate for relieving all pain wherever, whenever. He'll never know the lasting effect his image had on my career.

HOSPITALS HOUSED IN SMALL quarters have the advantage of frequent, incidental contact between colleagues. The hospital hallway is a common meeting place for physicians. On any given day, I encounter colleagues from a dozen or more different specialties. "How are you? By the way, I have a patient . . ." Known as a curbside consultation, this kind of informal interaction allows a physician to obtain information or advice directly from another physician about the management of a particular patient with a particular problem.

Hallway meetings tend to drive the hospital bureaucrats crazy since they're not set up for documentation and billing. But they are a way of life in hospitals and clinics. The curbside consult gives doctors a chance to find solutions more quickly and more easily. In our time of big-box hospitals, however, chance encounters are less frequent, and in this era of medicine—when seemingly every move requires copious documentation—they are generally frowned upon.

Hearing my name called out in the parking lot, I turned to see a neurosurgical colleague walking toward me with two women. After introductions, I learned that the younger of the two, Susan, had been a neurosurgery patient decades earlier, in her native California. Now a nurse at a nearby facility, Susan had come in, accompanied by her sister, for evaluation by my colleague.

When Susan was young, a shunt had been placed in her head. That's medical jargon for the solution to suffering from hydrocephalus—literally, "water head." A blockage prevented cerebrospinal fluid (CSF), the liquid shock absorber produced inside her brain, from flowing out and down along the spinal cord to be reabsorbed. Congenital hydrocephalus results from anatomical abnormalities, formed during the brain's development, that alter the free flow of CSF. Acquired hydrocephalus develops after trauma, especially when bleeding is involved (such as from a ruptured brain aneurysm), or from tumors obstructing the path of the fluid.

In the very young, hydrocephalus causes the head to grow abnormally large. With age the skull fully fuses, and with expansion of the head not possible, pressure builds—early on presenting as a headache or vomiting, but ultimately leading to death if not treated. A ventriculoperitoneal (VP) shunt is a plastic tube that diverts accumulating fluid in the brain to the belly. Anesthesia for patients with increased intracranial pressure (ICP) is tricky. Vomiting boosts the risk of aspiration pneumonitis, and inhaled anesthesia enhances blood flow to the brain, further adding to the ICP and increasing the risk of injury or death.

Susan had fared well until this moment, more than two decades later, when she began to suffer from headaches and sought help. During her appointment, her surgeon came in search of me.

"You're just the person I was looking for," he said. "It's not a shunt malfunction."

These shunts have a propensity to clog, necessitating a surgical revision. But the neurosurgeon had ruled out a malfunction, declaring that by his examination, the shunt was flowing well. Something else was causing Susan's pain.

I led Susan and her sister to a conference room within the anesthesia offices. They sat on a couch across from me. As I listened to Susan's story, I couldn't help but notice the magnitude of her pain. She occasionally winced, and the whole time her right eye squinted more than her left. After I asked where the pain was worst, Susan pointed to an area, and I leaned forward and tapped that spot on her head with my index finger. I felt the shunt tract, the tube coursing under her skin, off to one side, above and behind her ear. When I pressed on one particular spot, her head jolted back as I elicited a bolt of pain.

Susan suffered from a "trigger point"—technically a myofascial trigger point, more commonly known as a "muscle knot." Essentially, a small bundle of muscle had gone rogue and tightly contracted without relaxation, resulting in the stabbing pain that caused her to jump when I pushed on it. Many millions suffer every year from this poorly understood cause of pain.

I said to Susan: "Well, we can do this the official way, and I'll give you the number of our pain clinic, and they'll schedule an appointment and take care of it. Or we can forgo all the formality and I can treat it myself here and now."

Without a moment's hesitation, Susan implored me to end her pain.

A swab of alcohol, a syringe, a needle thinner than a common sewing needle, and an injection of just a bit of local anesthetic into the site of pain, the muscle knot, is all it took.

The change in Susan occurred instantly as I injected a contemporary derivative of cocaine, one that offers no central pleasure effect in the brain. The tension in her scalp and the wrinkles of her forehead visibly released. Her face flashed a momentary sense of disbelief; then a smile grew as the pain washed away.

Susan and her sister left, pain-free. Sometimes one injection isn't sufficient to ease the trigger point, and additional injections are necessary. But Susan never complained again after the first injection.

I'M NO CRUSADER. That honor belongs to colleagues and friends who have dedicated their careers to abolishing pain using new technology combined with old drugs, and old techniques combined with new drugs. These clinicians/researchers promoted the use of PCA (patient-controlled analgesia) using the standard narcotics linked with a computer-modulated delivery system, allowing patients to be masters of their own destiny. No more pushing a call button to summon a nurse.

The PCA system has been a great boon for patients. I mean no disrespect to nurses, for I marvel at their skills, their dedication, and their compassion. But if I order one milligram of morphine every two hours, the patient may not receive exactly that—the dose may be partially lost during the steps of drawing it into the syringe, proceeding to the room, de-airing the syringe, and administering it to the patient in pain. In addition, nurses have multiple patients with the same needs. Delay or dose inadequacy, though not intentional or immediately preventable, causes patients to suffer more than is necessary, and the cycle repeats. With PCA, the patient still pushes a button, but instead of the nurse's call button, it's a computer button that sends a prespecified dose of narcotic directly and instantaneously into the patient's IV. The computer is programmed to inject a specified number of doses within a specified time span, in order to prevent overdosing.

The cost for the right to pain relief is pleasure. Not too little, but rather too much. Narcotics induce pleasure by turning on dopamine, and inducing pleasure this way can ultimately lead to more pain, both physical and psychological. With time, increasing doses are needed to accomplish the same pain relief, leading to medication abuse and addiction. And these medications don't specifically and solely target the source of pain. Narcotics, or opioids—the terms are used interchangeably—possess whole-body effects that are not always desirable, such as lethargy, nausea, and constipation.

The seeds of narcotic addiction were sown long ago. The sixteenth-century physician Philippus Aureolus Theophrastus Bombastus von Hohenheim considered himself greater than the first-century Roman physician Celsus and thus took the name Paracelsus. In the 1500s, Paracelsus discovered that opium, insoluble in water, is soluble in alcohol. The early recipes for the mixture he named "laudanum"—from the Latin "to praise"—also included musk, saffron, cinnamon, and cloves. Laudanum was used for centuries for pain relief. It was even described as a sleep aid by Mary Shelley in *Frankenstein*.

During the Civil War, Dr. A. W. Chase—a traveling physician who published a lengthy list of medically related recipes—simplified laudanum to opium and alcohol only. In the Victorian era, the use of laudanum for recreational impairment became popular, just as the use of ether had been several decades earlier. In England, Godfrey's Cordial (a.k.a. "Mother's Friend") and in America, Mrs. Winslow's Soothing Syrup, were laudanum mixtures intended for children. Without oversight or regulation, this concoction contained more opium than was recommended for even a full-sized adult. Not surprisingly, reports of overdoses and deaths followed, eventually leading to the Harrison Act of 1914, which eliminated the ungoverned sale of over-the-counter narcotics.

The epidemic of narcotics-related deaths persists today with abundant back-alley, black-market sales. For example, heroin is easy and inexpensive to produce, creating expressways nicknamed "heroin highways" because a buy in certain parts of cities is as quick as the off and on ramps can be navigated. Imprudent prescriptions by physicians remain widespread. Medication orders include dose and frequency, both specified in excess of sensible limits. Over my career I have injected or prescribed literally gallons of morphine and its derivatives, almost all within the walls of a health care setting, since I rarely write prescriptions for addictive medications. I don't believe I have ever created an addict through my care, although it would be naïve of me to boast of complete innocence. I'm sure I've been the first to inject narcotics, albeit under anesthesia, into an addict-to-be.

In our present addiction craze, a seesaw battle seems to exist between overused, prescribed narcotics and illicit drugs. And this is a craze, an epidemic: every eleven minutes a death by overdose occurs. More people die by overdose each year than by falls, car accidents, or gunshots. Only when a celebrity dies does this epidemic reach the eyes of the media, where it is portrayed as a curse of fame, but soon after the VIP's death it is forgotten. As medical societies or the government move to curb narcotic prescriptions, street sales appear to pick up. Just saying no is not the answer. Overall, narcotic addiction is on the rise.

One radical thought is to limit the writing of narcotic prescriptions to specialties that are invested in pain relief—anesthesiology and pain management—and then make the practitioners in those specialties accountable for their prescriptions. Given the glut of required bureaucratic paperwork, physicians find the regulations for prescribing narcotic pain relief a time-consuming burden, with the result that it's easier to write a single prescription for ninety tabs intended to cover a three-month period than to write a new thirty-day prescription each month. Refills are not allowed on narcotic prescriptions. Again, the law of unintended consequences comes into play. Perhaps a policy that eases access through pharmacies but restricts the number of pills dispensed should be considered.

The goal of pain relief is obvious. Physicians treating pain should use techniques and medications that target the source of the pain while avoiding medications that have whole-body implications or that trigger the pleasure neurotransmitter dopamine, which potentially leads to dependence.

FINDING THE APPROPRIATE TREATMENT for the tethered boy in China nagged at me. My opportunity to shine came when I treated a man suffering a similar condition. Mike was born with pectus excavatum, a sunken chest. If he was lying flat, a cup of water would fill the hollow without flowing off his chest. Aside from its physical appearance, pectus excavatum can adversely affect breathing and the function of the heart, which is squeezed by the deformed sternum, the breastbone.

Few surgeries create as much postanesthesia pain as a pectus excavatum repair. All ribs are surgically detached from the breastbone, which is then fractured and re-formed with normal appearance. After the procedure, every breath and every movement of the torso hurts. The solution to Mike's pain wasn't to blast him with so much narcotic that he would pass the days obtunded, but to relieve the pain closer to its source, the chest.

With Mike's back to me, I inserted a needle into the space between his vertebrae at a point between his shoulder blades, passed a catheter through the needle, and placed the tip of the catheter in the space outside of his spinal cord but inside the spinal canal, giving him a thoracic epidural. Local anesthetic constantly bathes all the nerves emerging from the spinal cord at this level.

Visiting Mike after his surgery in the recovery room pleased me. He was awake and comfortable. In striving to eliminate pain for Mike, I hit a home run. Unlike that writhing boy in China, a few hours after surgery Mike sat in a bedside chair and spoke with me.

Not all cases end as well. Not all pain in all patients is eliminated. But that boy in China whose writhing I witnessed—a boy whose name I never learned— redirected my career and defined my goal: to wipe away all pain.

CHAPTER 12

A Brain Trapped in a Box

NOTHER STAGE IN MY PAIN RELIEF AWAKENING came during a vacation amid woods and water. Battling insomnia, I awoke hours before daybreak. My sleep for the night was over. With little else to do, and since staring at the ceiling seemed fruitless, I moved to the couch and turned on the TV. The choices were slim: watch the pitch for how to grow six-pack abs without breaking a sweat or succumb to purchasing a forever-sharp knife, able to cut nails and then slice a tomato.

I clicked on the only decent choice at the time: a documentary on the local public television station. The topic was communication in the cognitively impaired, people unable to speak or convey their thoughts to others. The scenes and settings of the show were so vivid that the faces and places remain indelible in my mind to this day. Despite repeated attempts to locate a reference for the film, I have found nothing. Sometimes I wonder whether that TV program was a figment from the dreamworld—a personal message intended solely for my growth as a doctor.

In the documentary, a man afflicted with cerebral palsy, his movements jerky and his speech low and drawled, detailed his life and his tribulations. I still see the image of him clearly, sitting in a small courtyard with a knife in his hand, whittling ever so slowly, his forearms moving like two gears in mesh—a short burst of motion, then a pause, his eyes wide open behind wide-rimmed glasses, his mouth constantly open, not closing even as he tried to speak. The scene changed to him sitting in front of his computer. He spoke about his condition. His newfound computer-generated voice verbalized the words he conceived but couldn't say. Unable to use his fingers, and instead gripping a pencil in each fist, he didn't tap; he *hammered* the letters of his keyboard and the icons on a touch screen that represented common phrases. The computer responded by vocalizing his thoughts. The man with CP spoke of being "a brain trapped in a box."

AS I WATCHED, I ENVISIONED myself during the many times I had stood next to a cart, looking at a patient in the recovery room or in the ICU and listening to moans I judged as too soft to indicate substantial pain. With the patient unable to advocate for himself, I routinely asked those present (families, nurses, other physicians): "Do you think it's pain?" The answer that almost always followed was "No."

A common cause of affliction in patients who are unable to advocate for themselves is cerebral palsy. The Centers for Disease Control and Prevention describes cerebral palsy as "a group of disorders that affect a person's ability to move and maintain balance and posture" and "the most common motor disability in childhood." Abnormal brain development or damage to the developing brain (caused by, for example, loss of oxygen supply) is a broad description that encompasses a multitude of potential causes of CP. Traumatic brain injury, brain tumors, seizures, and numerous other degenerative diseases might cause an inability to move smoothly and, more important, to communicate. With adulthood, stroke and burst cerebral aneurysms enter as causes of brain injury.

The night of my insomnia, my revelation placed the onus for the relief of pain on the shoulders of the person holding the ultimate responsibility: me.

One patient in particular came to mind: David, a young man with a body twisted tight as a result of CP. His father was an acquaintance from my neighborhood and knew that I was an anesthesiologist at the hospital of his son's scheduled surgery. I don't recall the cause of David's CP—whether he'd been born too early, or there had been an infection during his development, or birth trauma had led to a loss of vital oxygen. But David's newborn brain never overcame the hypoxic insult he suffered, never healed, and failed to develop normally. Now, deep into his second decade, David's brain fired in uncoordinated bursts, prohibiting the fluid motion necessary to accomplish the complex tasks we take for granted, such as tying a shoe or lifting a fork to the mouth. Or speaking.

After what I believed to be a minor surgical procedure, David lay on the cart, gently moaning. His movements—those he was capable of making—were slow and ratchet-like, leaving him unable to walk, write, or speak. On his cart, he assumed a birdlike posture. He lay on his back with his arms flat on the bed and slightly abducted (extended out from his chest) bringing his elbows to rest on the sheet six inches or so from his ribs. David's elbows were flexed, tightly raising his forearms toward his head, his hands coming to rest about a foot to the side of each of his ears. He wore a neat beard, evidence of someone having provided loving care. His head was turned to one side, and his mouth was open like that of a chirping bird, forming an almost "O," his tongue visible.

His moan came in soft breaks. His hands tremored slightly and briefly, and then a short rest ensued. He tried to turn his head to the opposite side but didn't quite make it and returned to the side where I stood. His eyes were fixed on me. Then the cycle repeated. (These patients are unable to smile or grimace, so the FACES scale of pain assessment is not accurate.)

I spoke to David's parents in the waiting room, informing them that from my anesthesia perspective, all had gone well. I escorted them to their son's side in the recovery room with the intention of gaining their insight about David. I needed some direction regarding additional pain relief needs. I had provided a dose of narcotic pain reliever and thought that was sufficient, given the nature of the surgery. Upon seeing his parents, and after a deep inward breath, David appeared to try to force out his thoughts, but to no avail; his voice uttered deep grunts, and his tremor grew more pronounced.

"Do you think he's in pain? I don't want him to hurt. If you think he's in pain, I'll give him more medication. I only have this snapshot of him to go on. Tell me if you think he needs more medication."

My window of observation was too narrow for me to make a call. Surely, his parents must know better than I could.

Their response was no different from that of the families of all the other Davids before, and I'd cared for many hundreds with this affliction. His parents explained that he always reacted that way. He was excited to see them. How they knew that, I couldn't tell. I observed their son awake for a total of no more than a few minutes on either side of my anesthesia.

This behavior, for me, was typical of CP patients in the recovery room after a not-too-invasive surgical procedure. I could never communicate with them, I couldn't tell how much discomfort they were in, and I guess I assumed their actions would be more demonstrative if they were hurting.

WITH MY SLEEPLESS NIGHT, and with my pain relief epiphany, all that changed. It dawned on me that I had undertreated David.

I awakened to a world I had misread for far too long. I had failed, until now, to consider that a person so physically impaired and unable to speak might be able to absorb the surrounding environment. The man in the documentary stunned me. An intellect, a poet, a writer, gainfully employed—and all because his parents had believed in him, insisted on educating him, and mainstreamed him through school, and then technology had caught up, allowing him to attend college and reach outside the body that trapped him.

My mind fast-forwarded to the moaning I had heard in the recovery room emanating from those unable to communicate their thoughts. I had misread their hurt, their anguish, as simply confusion brought on by the unfamiliar setting. I felt the sum of their pain as I realized that *I* was the one who had been unable to communicate. Whether their ailments were caused by cerebral palsy, genetic alterations, brain tumors, vascular accidents, or traumatic brain injury, hundreds to thousands of my patients existed beyond communication. How ignorant I had been not to listen to those soft moans! I hadn't heard their attempts to tell me they were in pain.

Those taunts as a kid, after a botched play or a stumble on playgrounds and sporting fields: "You palsy!" It was right there. A failure of movement. It was right before me. I had heard it but hadn't listened. I had never bothered to look up the definition. "Palsy": "paralysis accompanied by involuntary tremors." The definition mentions nothing of failed receptive capability, the ability to hear and to understand.

On my couch that night, I remembered a grade school classmate's sister—a girl with CP—and her scissors walk, stiff legs never relaxed, struggling to and from school every day. I recalled the cruelty of the boys lobbing piercing comments at her. Her body prevented retorts, but she might have heard every word and understood everything. I remember the toes of her shoes worn from dragging the pavement—a stark contrast to her wide smile.

I found myself enclosed in a cloud of guilt. That I wasn't the abuser didn't exonerate me. I hadn't defended her. In the dark, lit only by the TV screen, I decided I had to become a stronger advocate for any cognitively impaired person—regardless of the cause of brain injury—coming to my care. It's left to me now to defend my grade school classmate through my patients of today.

MY AWAKENING CAME ABRUPTLY and in the dark. *A brain trapped inside a box.* Though able to perceive everything about him normally, he could not express himself.

After nearly fifteen thousand cases, I finally asked this question: Can cognitive ability (the means to perceive, to understand, the receptive capability) be separated from expressive ability (voicing thoughts and feelings)? Until then, I had assumed that if the reaction was not possible, neither was the reception.

Ashamed to admit it, I was halfway into my career before I actively pursued speaking with noncommunicative patients as if they were typical for their age and possessed normal thought process, regardless of how densely involved I considered their CP to be. Until proven otherwise, that's how I proceed now. I always explain with great care every move I make, ensuring that there are no surprises.

The weight of this revelation led me to embrace a new nonnarcotic, nonaddictive medication for pain relief given by IV. The value of the drug, ketorolac—a pharmaceutical offering that, at the time, had only recently been introduced—is not so much that it is magnitudes more potent than Motrin (the widely used anti-inflammatory pain reliever available at the local drugstore), but that it is formulated as injectable, able to be given via an IV. Finally, here was a nonnarcotic pain reliever (with no risk of respiratory depression) for which swallowing was not necessary—the only available parenteral nonnarcotic anti-inflammatory that could provide strong pain relief.

My chance to step up materialized when another boy with CP came for tendon releases of his legs. Muscles in constant contraction cause tight joints with locked knees and legs unable to spread wide, making for difficult hygienic care. Life becomes so much better for parents, caregivers, and the patients themselves when the joints are relaxed, allowing for easier transfer from bed to chair and improved cleansing and dressing. Creating movable joints isn't very complicated or intensely invasive, but it does necessitate incisions and cutting of tendons and requires pain relief following emergence from anesthesia.

After his procedure, I stood beside this boy in the

recovery room. My emotions swung between satisfaction and anger. He, too, lay in that classic birdlike posture, his hands to the side of each ear. Except he wasn't moaning. He was quiet. He was comfortable. He was pain-free. The joy at my ability to safely alleviate his pain was tempered by the recognition that for years, I had failed to understand or treat pain in my noncommunicative patients. I had failed so many patients.

Those unable to speak for themselves, like the man in the documentary, had no advocates for comfort. Their families were either incapable of understanding them or, like me, fearful of overmedicating. I changed from being a reactive analgesic administrator—depending on others beyond the patient to provide guidance for pain relief—to an activist, making pain relief my decision, with every intent to prevent as much moaning as possible. Instead of steering clear of overdosing, I came to define its limits with greater precision. It was all thanks to that brave man with CP who rose tall in his wheelchair and, as he presented his speech, enlightened me.

Sometimes I wonder how future generations will judge medicine, surgery, anesthesiology, and me and my actions as a physician. Will they view my career as barbaric, as I view the barber-surgeons of the nineteenth century?

CHAPTER 13

See One, Do One, Teach One

I N THE MIDDLE OF THE OR, IN THE MIDDLE OF A CASE, a colleague informed me that my wife was trying to reach me. On the wall behind my anesthesia cart, the light on the wall phone was blinking. I lifted the receiver and cautiously said hello. I knew she wouldn't be calling to remind me of some event taking place that evening, or to ask me to pick up something on the way home.

"Nathan's throwing up pennies."

Standing in a procedure room nearly twenty miles from home, I was asked for a medical opinion on my own child, who was then just a toddler. Considering the size of a penny, I knew swallowing one wasn't life threatening.

"Call me back when he's vomiting quarters."

With my response, all eyes in that procedure room turned to me. My reply may have sounded cold and uncaring, but after many years of a dedicated relationship, my wife and I understood and trusted each other.

Three minutes after first meeting, I ask patients, spouses, parents, children, and any others present to trust me with their own life or the life of a loved one. I expect that trust, and I need it.

The expectation of expertise comes with the approval of medical-center staff privileges. Becoming a staff anesthesiologist, or a staff member in any specialty, for that matter, requires holding a state license and usually demands further credentials, including board certification. For anesthesiology, completing an approved training program, logging the necessary number and variety of cases for the specialty, and passing a written examination are followed by the final hurdle: successfully mastering an oral exam that consists of two thirty-minute grillings by seasoned anesthesiologists on any topic in the specialty. I've spoken with many examiners, and most agree that the grade is determined within the initial ten minutes of the oral presentation.

Snap decisions are a reality. My patients and their families make them time and time again every clinical day of my life. And they do so without question. That surprises me, because I can count on one hand the number of times in my entire career that I have been asked to present my credentials prior to my care. It is wise for patients to research their physicians. I believe such investigation is more critical for physicians practicing outside the realm of credentialed facilities. I would prefer to be asked about my board certification and recredentialing.

Trust is elevated to another level in teaching hospitals. I train residents to become tomorrow's anesthesiologists. I teach fellows, residents who have completed their training, homing in on expertise in pediatric anesthesiology. I supervise these trainees as they administer anesthesia.

A MOTHER CALLED ME about her daughter and an upcoming surgery. She was concerned about the anesthesia care for a fairly routine outpatient procedure. But when it's your child, there are no "routine" procedures. She did what every patient or family should. She asked around to find the names of anesthesiologists referred by friends and caregivers.

That she wanted more information about my anesthesia plan didn't surprise me. What does surprise me is how seldom I field calls like this. Within our department, a system is in place to answer phone calls and questions, with an assigned anesthesiologist scheduled to field questions every day. For me, this was not that day. This mother sought me out specifically.

Our office secretary paged me concerning an outside phone call. Just outside the OR suites, at the intersection of two great halls, on a wall phone down the least used section of hall, I responded: "Hi. This is Doctor Jay."

"Hello. Thank you for talking with me," the mother said.

Phone calls are similar to face-to-face time: too short, except with no eye contact. Peppering the conversation with a few pleasantries accomplishes two goals. First, the parent or patient is put at ease and encouraged to speak unrehearsed. Second, time takes on a different dimension. The time spent in conversation feels longer than it is. The person I'm speaking with believes that time is not the driving force, that I have their best interest at heart. I mine the information I'm searching for, in addition to some helpful side information that will allow me to form a bond more easily, and I do so without seeming to interrogate the person.

The mother eventually came around to asking this question: "Can we talk about anesthesia for my daughter?" She was calling to ask specifically if I would care for her daughter.

"For curiosity and nothing more, how did you get my name?"

She mentioned that she had asked around, and my

name kept surfacing. That she needed to come to my hospital wasn't unusual, since I practice at the most complete pediatric hospital in the region. We offer care for every conceivable childhood health care problem, and care for children from around the world. What I found surprising was that she asked for *me*.

She buffed my ego. Perhaps the person she had spoken to knew only my name, or the recommender knew nothing of my qualifications or ability. But I chose not to pursue it further and just accept the flattery. My experience has grown over the years in practice, and I have sat on committees, both regional and national. Still, most requests have come from people I knew. This was an uncommon cold call, and request.

I patiently awaited the real reason that she had called me. I suspected we were edging toward ethical quicksand, and I was in no rush to get there. Soon enough, after she asked whether I was available on the scheduled date and I assured her I would make myself available, she popped the real question: "And you'll do the anesthesia yourself? I don't want any residents involved." She wanted me, alone, to touch her daughter. Trainees, if present, were to sit in the corner and observe.

A giant, circular argument opened, one that never comes to rest. It's the argument of teacher versus student.

I'm not sure if there is a service in the hospital that doesn't frequently hear the request for only the attending surgeon to hold the scalpel. For only the attending pediatrician to start an IV or draw the blood. For only the attending anesthesiologist to anesthetize and intubate the child.

The saying in medicine is: *See one, do one, teach one.* Before any of this, learning begins with reading. Observation follows, but expertise doesn't come with watching. Observation yields only so much information. Hands must be laid upon the human body to move beyond the wisdom that book learning imparts.

A pianist doesn't master a piece by Bach or Beethoven by watching a video over and over. Those fingers must touch the piano keys, just as the physician's hands must touch the body, to learn. Inexperienced hands have to be given the chance to palpate what's normal and, equally important, to distinguish what's not.

What of the parents requesting—or worse, demanding—that only the attending doctor, and not the resident, touch their child? The response varies from doctor to doctor, hospital to hospital. Many oblige, responding simply: "Sure." Behind closed doors, nobody but those in the room knows who actually touches the patient. Nobody in the room except the one questioned might know of the request.

Since fewer than twenty percent of all hospitals are teaching hospitals (that is, hospitals with resident training programs) and many of these teaching hospitals have limited residency programs covering only a few specialties, the issue of requesting that no residents be involved arises in only a minority of health care settings. But in those settings, it surfaces often. Patients and families have options. A community hospital might be a better alternative for attending-only care. Regardless of the setting, care is by teams consisting of physician assistants, technicians, and advanced practice nurses. The concern that no residents practice on someone is couched in a different cloak, but it still remains. Physician assistants and advanced practice nurses do provide patient care.

The tipping point motivating patients to seek out teaching hospitals—the university systems and medical schools—is that these are considered to be leaders in health care, the research centers expanding the envelope of medicine. Experimental procedures and the newest techniques developed in these hospitals might not be available elsewhere. The academic setting represents the cutting edge of health care, where physicians aren't under the continuous pressure to generate bills and are allowed to apply creativity to the practice of medicine through research and clinical studies that consume both time and money.

Nearly every hospital nationally ranked for any specialty is a teaching hospital. Seven of the top ten hospitals surveyed in one urban area are teaching hospitals.

Patients entering these halls of medical learning should understand from the outset that trainees are a fundamental part of the process.

My response to attending-only requests is rote: "No. Don't ask me to change my practice." My care is built around a training program. Asking for attending-only care alters my process, and that increases the likelihood of mistakes or missteps. If blame for the failure of care is to be placed for any reason, blame me; but don't ask me to change my process. Besides, the trainees' hands are an extension of mine.

My trainees evaluate and examine the patient. They perform the same rituals that I have followed for many years. After the assessment is completed, a plan is developed. Before this plan is put into action, the resident and I discuss everything. I confirm the history, physical, and all data obtained, then approve the plan or offer an alternative. In my system, two physicians evaluate the patient. Why not take both of us?

To this particular mother, I said: "Please don't ask me that. You know this is a training program. And I'm telling you, I take full responsibility for your daughter's outcome. If you have a problem, you bring it to me. You need to trust me. I won't alter the process for you, but I'll tell you the surest way that I can provide your daughter's best outcome is to let me do it my way."

Her response was delayed. I heard her take a breath; then she acquiesced.

On the day of the procedure, I met her with her daughter for the first time. I reiterated my intentions. I reconfirmed that if there were any problems with the anesthesia care, they were mine and solely mine. And then I added: "There will be no problems."

Her daughter navigated the system without issue.

How far should a resident be allowed to go in the learning process? A resident who is unable to demonstrate an understanding of the health condition at hand or the necessary technique of care is bumped to the side and offered a seat as a spectator. Once an understanding of the condition and the required care is clearly demonstrated, the limit is to never, ever allow injury to a patient. Hard stops—cases where I prevent the resident from trying—occur, among other times, after three missed attempts at an IV, or two attempts at inserting the breathing tube. "Do no harm" is a constant voice in my physician mind.

I WALKED INTO MY patient's room and wandered into a maelstrom. The sliding door was open, but the privacy curtain was closed. As I swept the curtain aside, the parents' tension was unmistakable.

A little boy, about three years old, was jumping up

and down on the cart. His father stood next to him with both hands on his son, preventing him from falling. That alleviated my fear that this boy would leap from the gurney.

Behind the father, three chairs lined the wall. Nearest me, at the entrance, were two industrial-style armchairs—metal framed and with stout, straight backs and blue faux-leather cushions that coordinated with the décor of the curtains. Intended to provide a homey feeling while suitable for heavy use, the hard, plastic arms don't quite measure up. Furthest from the door and angled in the corner sat a light cranberry-colored leather chair with soft cushions. The reclining feature does succeed in providing comfort, both to the appearance of the room, and to whoever sits in it. Not infrequently, I'll catch a dad napping there. Mostly, I find moms holding and snuggling a child, providing comfort.

The space between my patient and me was occupied by a parent. In my experience, that space is most frequently filled by a mother, simply because eight out of ten times, it's mom who provides for the health care needs of a child. This space existing in pediatric health care is not normally present in the adult health care world, where families sit (or stand) to the side.

Usually this space is like calm water, easy to navigate. Parents, although concerned and nervous, are also anxious to be of help and provide every bit of information necessary, allowing me to develop a safe and effective treatment plan for their child. Usually I ask the mother most, if not all, of the questions. Mothers provide for the majority of their children's health care. Fathers—myself included—more often than not don't know all the answers.

"Does your child have any allergies?"

"Uh . . ."

Once this space has been navigated, it's on to the child so that I can complete my examination while establishing a relationship.

Occasionally this space is not so calm. My hospital is a tertiary-care, pediatric hospital, meaning we are a final destination for the patient and, more, the parents. Community hospitals might not provide the necessary care, or the care in the community might fail to alleviate the problem. An unsuccessful or unpleasant experience with another system leaves patients, parents, and families not only anxious, but sometimes also angry or distressed. This situation is tricky to navigate, requiring insight into the reason for the transfer of care and the cause of the discontent, while demonstrating an ability to provide the appropriate care, assuring all that we can do better and showing compassion not just for the child, but for the parents.

This bouncing boy's mother sat with her back

straight, her shoulders high, and her arms folded tightly across her chest. This wasn't the appearance of an anxious, unknowing mom. In this setting, I'm accustomed to parents posing uncomfortably—sitting on the edge of the chair, with hands tucked under the backs of their knees—attempting not to appear too tense. No, this mom represented a funnel cloud ready to touch down, like a spring coiled tightly and ready to release all its energy with the slightest trigger. This mom was ready to strike, and I stood in her path.

Her child had come in for a procedure on the penis— another one. In his three-year life, this was to be his third attempt at penile beautification, the first two done elsewhere, and this mom was clearly frustrated. Her frustration loaded the spring, and anybody who made the slightest out-of-step move was a potential trigger. As always, on entering, I introduced myself. The little boy's father responded, while the mother sat still and quiet. Her gaze shot through me; her angst froze me.

It amazes me how much penile pathology exists in young boys; still, I find it equally amazing, if not more so, how obsessed parents are with their sons' penises— as if their boys would be condemned to a life of ridicule and assured of failure if any aspect of this mini-organ was just a little bit off. Fortunately for me, I am not a pediatric urologist. (Adult urologists have plenty of nicknames, from "dick docs" to "stream team" to "prick

plumbers"; my favorite is "wee-wee whackers.") The
pediatric urologist must deal with many preoperative
and postoperative visits, and the litany of questions and
concerns that come with each. As a pediatric anesthesiologist, my contact time is normally limited to the day
of the procedure.

RECOVERING FROM THE INTENSITY of the mom's
gaze, I asked all the usual questions. The father answered
while the mother sat, simmering quietly in the corner.
I examined the boy, and then I started to discuss my
proposed anesthesia care with his parents. I noticed that
during his previous surgeries, he had not received caudal blocks for pain relief. I was instantly reminded of
Casey, the master surgeon who is one of my legends.

Casey's patients were primarily children with penile
issues, and the patients' parents raved to him about their
sons' comfort lasting long beyond the recovery room,
long after discharge from the hospital and their return
home. The caudal block, a single injection of local anesthesia medication placed while under anesthesia (no discomfort) upon the nerves at the base of the tailbone lasts
six hours or more. One particular block lasted so long
that the first pain reliever was not necessary until sixteen hours after surgery.

Casey had sought me out. "More satisfied parents!

More caudals!" He wanted every patient that might benefit from the block to receive one. I followed his guidance, and I was proud that these children not only left my care pain-free; they also left the hospital and arrived home in comfort.

But in this little boy's mother on this day, my recommendation of a caudal block triggered the spring's release. Angst or anger—I couldn't differentiate—overflowed. The reason for her insistence that there be no block might have been a failed epidural block during her labor and delivery, or the common misconception that caudal blocks cause permanent paralysis. Untrue. If she'd had confidence in me before I mentioned the caudal block, I risked losing it by insisting that her son receive one. I chose another tack.

During this time, the little boy continued to bounce up and down in the safety of his father's arms, unaware of any controversy.

I was convinced I could provide better pain relief after this procedure than the boy had received in the past. My easiest option was to leave well enough alone, drop the topic, and proceed without the block. But my first responsibility is not to the parents, but to the patient. I wanted this boy to receive the safest, smoothest, and most pain-free anesthesia possible. To return him to his parents awake and alert and pain-free was my goal. I pushed forward.

My challenge in this case was determining how far to go in insisting that the parents accept my plan for care, and for pain relief—care above and beyond that of the child's previous procedures—even in the face of doubts. Past the "See one," continuing beyond "Do one," and now sailing past "Teach one," I was moving beyond the two thousand caudals performed and not one substantial complication. I proceeded.

"This is one of the few times that I can remove myself as a physician and tell you, as a father, what I would do. I must tell you that, without any second thoughts, I would never let my son have this surgical procedure without a caudal block."

"Really?" the father responded. "But is it safe?"

"I have performed thousands of these blocks without a single significant complication. If there has ever been a problem, it's always been that the medication didn't reach the exact area it needed to work properly. If this happens, I will give your son the medication I would have used if I had not performed the block."

"Hmmm." A pause. "Well, I think we should try it."

Before the words were fully off the father's tongue, the mother jumped forward from her seat. Her arms flew out and she glared at me with a look between accusation and total disdain.

Then she said it: "I've heard these blocks can cause permanent injury."

"Never in my experience." I have never cared for a patient who suffered permanent neurological injury from any nerve block. And I repeated that of all the blocks I have performed, I've never had any complications besides failure of the medication to relieve pain.

"I will do what you want," I continued, "but I strongly recommend this block. This is the single best manner that I can provide pain relief for your son."

"Let's go for it," the father declared.

The response was greater than a shift of the San Andreas Fault.

The boy's mother took a few steps toward the foot of the cart. She looked at her husband for a moment, then at her son, then back at her husband. Her right arm flew out in front of her, and she wagged her extended index finger directly at her husband and declared: "If anything happens, I want you to know it will be on *your* shoulders!"

Attempting to leave on a high and confident note, I said: "Trust me. We'll do just fine."

The pressure was on me to deliver. I walked past the curtain and, before turning up the hall, slid the glass door shut. I didn't want to hear the discussion that took place once they were alone.

After entering my pre-anesthesia note for the little boy, I wondered for the briefest of moments whether I had overstepped my bounds and pushed too hard for the caudal block.

I completed a final review of the record, checked the consent, and walked back to the boy's room. I asked if there were any final questions. The mother never returned to her chair and stood silently scowling in the corner. The father responded that everything was fine. I lifted the boy off the cart and carried him back to the OR.

We walked through the double doors and I let out a long "Yesss!" I had taken the boy from his mother, and he wasn't crying. This was good. I talked to him the whole time about favorite TV shows and books. I started reciting *Green Eggs and Ham*.

As I placed the bubblegum-scented mask to his face, I kept speaking. I told him he might feel *tingly* and possibly even *giggly*, since he was breathing laughing gas. In a few seconds he was quietly anesthetized.

The IV went in uneventfully; then I turned him on his side, while anesthetized. I carefully examined his back to make sure a caudal block was possible, that there were no congenital defects of the spine that would preclude my placing a needle into the epidural space and injecting local anesthetic. His anatomy was perfect.

I looked at my anesthesia resident. Knowing what had transpired, she declined to place the block. At first, this disturbed me. How else can one learn? Then I considered that pediatric anesthesia wasn't her interest, and that she wouldn't likely ever perform this block in her practice. I had no problem placing the block. The needle slid in and the local anesthetic flowed smoothly. The case proceeded uneventfully. The surgeon did a marvelous job. This penis would heal just fine, with very little scarring; his would be a penis that any parent would be proud of.

Most important to me, this boy awoke pain-free, and I transferred him to the recovery area.

With a full schedule, I hustled on to my next patient. I stepped up to the gurney to a woman sitting on the edge. I looked to the floor and watched as perspiration dripped from her feet, which dangled in the air, puddles forming under each that coalesced in my presence. This woman suffered from hyperhidrosis, the real disease, which is nothing like the claims of underarm wetness. The perspiration in true hyperhidrosis is so severe that hugs feel like embracing a soggy towel, a handshake feels like grabbing a soaked sponge, and beads of sweat grow spontaneously. Clothes and shoes risk being ruined in one wearing.

Botox, famous for its ability to prevent facial wrinkles, offers a therapeutic option, although expensive. The sites of greatest sweat production include the hands, feet, and armpits, as well as the gluteal fold between the buttocks. The many, widely spread nuisance glands causing the problem are injected with hundreds of minute amounts of Botox. The multitude of injections warrants anesthesia. The case was easy for me: an IV, milk of amnesia (propofol) for induction, and sevoflurane for maintenance. Soon I transferred the woman to the recovery room with a thousand, perhaps more, minuscule drops of blood visible. She awoke grateful.

Next to her in the recovery room, my young patient from the earlier case sat on his gurney enjoying a popsicle. His mother sat at the foot of the cart, his father not in sight.

"Well?" I said.

"Well what?" Mom responded.

"What do you think?"

"He's OK."

I needed to control myself. "OK? He's great. It went perfect."

The next day, I was shocked to learn that this mother had mentioned that their hospital experience had been the best they'd ever had.

THE REWARDS OF BEING a physician come to me from many directions. Most, but not all, come from providing patient care and knowing that I made a difference. Others come unexpectedly. A friend and previous colleague called to ask about a friend of his whose child had "special needs" and required anesthesia. I told him to have her call me. She did, and I must have allayed her fears and concerns. Shortly after, he sent me a letter referring to me as a "mensch." An honor indeed.

Some of my fondest memories come from watching others master the skills and techniques that I have taught them. I received a call one morning from a young doctor I'd trained, who was now practicing hundreds of miles away.

"You saved a life during the night," he said.

"Wonderful. How did I do that?"

He explained that a newborn had been transferred to his NICU. The infant was in respiratory failure and had a tiny mandible. This meant he was not able to breathe well. The neonatologists had tried many times but couldn't intubate him. They couldn't see the vocal cords. "They called me, and when I saw the baby, I remembered how you taught me the retromolar approach." This is a maneuver on intubation to visualize the difficult airway. "I saw the vocal cords on the first try, and the tube went in without trouble. You saved that baby."

That phone call is my favorite teaching experience. It was a pat on the back, the reward for teaching, the fulfillment of the oath of Hippocrates by adding to someone else's expertise.

CHAPTER 14

Reentry

OST OF THE ANESTHETICS ADMINISTERED TODAY continue to utilize an inhaled potent ether gas. For the patient, reentry to awareness entails reversing the chemical coma. I turn the vaporizer dial two clicks to the left, cutting the flow of the volatile anesthetic agent. The gas is literally breathed off mostly unchanged, and when its concentration in the body drops to sufficiently low levels, the patient emerges from anesthesia. This process takes place more slowly and is more readily counted in minutes compared to the seconds it takes for anesthesia induction. Akinesia (paralysis), if used, requires a reversal drug—hopefully, avoiding a syringe swap.

Patients, and especially surgeons, want an anesthesia button that is pushed once at the start of a procedure to

instantly induce anesthesia and, at the end of the procedure, is pushed one time again to turn the anesthesia off.
We're pretty close to instant anesthesia induction, but
instant anesthesia emergence is still some way off.

Dividing the aims of anesthesia into the Five A's
allows anesthesiologists to expedite the on/off goal, and
to improve patient outcome and satisfaction. Treatment
plans that avoid using volatile anesthesia gases, especially for procedures that do not invade the abdomen or
chest, might shave some time off the emergence process.
Avoiding the ethers might have other benefits. A person
with persistent nausea and vomiting after previous
gas-based anesthesia might benefit from not breathing a
volatile agent, a known instigator.

Using combinations of drugs that alter anxiety,
amnesia, and analgesia can achieve the desired anesthesia effect but requires multiple drugs and exposes the
complexities of consciousness. We form memories while
asleep, but not while anesthetized by volatile gas, and
we lose awareness while we're awake during daydreams,
as well as under the influence of azepams. There's always
a risk that a patient will retain awareness somewhere
along the process of anesthesia.

Not uncommonly, I'm urged by patients and families to "not use too much," "to go light," to provide
"twilight," or to avoid the gas anesthetics. Recent

damning press has reported on cognitive dysfunction after gas anesthesia. Scientific studies to date don't support this conclusion. I liken the control of anesthesia to light switches. I'm most comfortable assuming all control of the patient. This is the toggle switch approach: either on or off, with no in-between. When I use a volatile gas, the patient is completely anesthetized, I'm in total control, and I'm able to deliver the patient back to consciousness with the lowest risk for complications.

Veering from this approach, altering the depth of altered senses by decreasing the dose of anesthesia drugs administered simulates a dimmer switch. I relinquish total control of the patient. Neither the patient nor the proceduralist nor I possesses total control. As the dial on the switch turns, and it is not always turned by my fingers, the amount of drug needed changes, and at some point the patient enters the state of full anesthesia—a point not easily recognizable. Complications and even death result from failures to spot and promptly treat drifting blood pressures, aspiration of gastric contents through an airway not able to close, airway obstruction, or the cessation of breathing. And such "reduced" sedation is often provided without the presence of an anesthesiologist.

Regardless of the type of anesthesia administered, I

believe that assigning complete control of the patient to me ensures the safest outcome.

THE TREND AWAY FROM surgery and toward minimally invasive, interventional care continues to grow. Replacing leaky or constricted heart valves is moving from open-heart surgeries to the cardiac cath lab, clipping cerebral aneurysms through craniotomies is shifting to the radiology suite for coiling, snaring cancerous polyps in the colon is transitioning from abdominal surgery into the endoscopy lab. And with these changes, the needs for anesthesia are changing, as are anesthesia techniques. The pre-anesthesia discussion between patient and anesthesiologist—what's wanted, what's needed, what's advised—is as important as ever.

Toward the goal of minimizing the effect of surgical invasions on the patient, arguably the most significant pharmacological advance in anesthesia in the last forty years came from a new and unique sedative, not a tweaked chemical formulation of an existing drug. Propofol, affectionately known as "milk of amnesia," is poorly soluble in most carriers; it is dissolved in a lipid carrier that gives it an opaque white color. Its short action, five minutes, provides for a quick return to awareness and makes it ideal for sedation as a continuous IV infusion. Propofol provides two of the Five A's:

anxiolysis and amnesia. It provides no analgesia and actually burns on injection. Painful procedures require additional medications and techniques.

Despite its short action, propofol is a dangerous drug, as evidenced by Michael Jackson's death by overdose, a reminder that vigilance when altering alertness remains a necessity. The biggest advantage of this drug is that it speeds reentry. All the other drugs I use to treat anxiety and amnesia entail a hangover period, in which the patient feels altered and unable to perform normal tasks. With propofol, there is no hangover. Propofol allows consciousness to return rapidly and intact, with the added benefit of reducing postanesthesia nausea and vomiting.

MY SON'S REENTRY TAUGHT me a valuable lesson. He received remarkable anesthesia care and emerged intact and well. But not long after, he confided to me that he had been in pain after the procedure. In my experience, craniotomies surprisingly lack pain, even though bone (the skull) is being cut—normally a cause of significant pain. He enlightened me. Most of the craniotomy patients I've cared for had brain surgery toward the rear of the head. My son's surgery required a frontal approach that cut through the muscle of the forehead, the temporalis. Every facial grimace included contraction of that muscle, and thus pain. Narcotics could provide the needed analgesia, but at a price of possible sedation, an undesired side effect possibly masking assessment of brain function.

Karl Koller, an Austrian ophthalmologist, was an early adopter of local anesthesia. In 1884 he experimented with injecting cocaine into his own eye. I've injected lidocaine into my own lacerations to lessen the pain of wound suturing. The pain relief team responded to my son's pain by injecting a local anesthetic near the nerve that innervates the muscle in question. The result was spectacular.

The actual advance that made prolonged pain relief by injected local anesthetics successful came by way of placing a needle accurately near a nerve. Originating as a method to detect submarines during wartime nearly a century ago, sonar (originally named for "sound navigation ranging") underwent a variety of technological advances that led to a probe the size of a pack of playing cards. When held in the hand and slid on the skin over a nerve, it creates a sound-generated video showing a needle's position and demonstrating local anesthesia during injection bathing the nerve deep in the body. There is no nerve that can't be reached with an ultrasound machine. Using ultrasound, a well-trained anesthesiologist can provide pain relief to some of the most difficult patients—for example, those with cancer in the abdomen and those with incapacitating back pain.

There is a deep-seated belief that anesthesia steals something from the patient's mind. Reuniting family with patients in the recovery room too early only strengthens this erroneous belief. The gas anesthesia agents wear off in a toe-to-head pattern. The patient is able to move and even respond to commands very soon after surgery, but recognition requires the highest brain function and is the last to return during reentry. A patient's inability to recognize family can be frightening.

One uncommon source of distress during reentry is delirium. Even while appearing fully awake, a patient emerging from anesthesia may not be oriented to person, place, or time. The cure is not to administer a "tincture of time" to allow the delirium to resolve, but to sedate the patient to enable slow reentry. After a brief sleep, the patient emerges peacefully, intact, and able to recognize family and friends.

Reentry is just as magical as induction. My faith is a necessity in my career as an anesthesiologist. After more than thirty years of practicing, I'm no closer to explaining the mechanism by which the gas I provide anesthetizes, and I'm as baffled as ever about how the drugs I administer selectively alter memory. When I emerged from my own experience of anesthesia, what

I needed most was to see my wife, her face in conflict, concern for my well-being obvious, and overwhelmed by that gentle smile of love that provides more comfort than any medication. Observing my patients happily reuniting with their spouses, parents, and loved ones, and in better condition than on arrival, fills me with an enduring joy.

CHAPTER 15

Safe Travels

———

NICK, THE UNFORTUNATE HALF OF AN UNEQUAL twin gestation, was born with the VACTERL association, a constellation of congenital defects that tend to occur together. The defects include anomalies of the spine, an imperforate anus, heart malformations, disruption in the path for eating or connection of that path to the trachea, abnormal kidneys, and deformed arms. Nick suffered from the life-altering radial hypoplasia—a misshapen forearm with no thumb, creating a minimally functional hand—and a life-threatening congenital heart defect.

Nick's first anesthesia experience came shortly after birth for the repair of a tracheoesophageal fistula, which allowed fluid to pour from his stomach directly into his airway, avoiding the cough protective reflex and flooding his lungs. He returned for anesthesia many times thereafter. Pediatric anesthesiologists train for these neonatal cases, which involve multiple organs and demand precision.

I can't remember the first time I provided Nick's anesthesia care. It was early in his life, and I recall a tracheostomy in place—his voice box narrowed below his low-functioning vocal cords, requiring a tube in his neck that allowed free unobstructed breathing. After the procedure, I talked with his parents, both in recovery and later in his room. They asked if I would be able to care for Nick when he returned for his next procedure in a few months. "I'd be honored," I said. "Just give me a call." I believe Nick's parents felt reassured that someone was concerned about Nick as more than just a patient or condition. Anesthesiologists often limit their discussions to the case or the condition. I spoke with Nick's parents about their son.

A couple of times each year, I would receive a phone call from Nick's father—"Hey, Doctor Jay, are you available?"—and Nick would undergo another procedure. After every anesthesia, when I spoke with Nick's parents, it was his mom who did most of the talking, so I found it curious that his father always made the phone call. As Nick grew, so did the number of times I cared for him. My response to Nick's father was always: "When and where?" followed by "I'll be there." I would mark my calendar for the date.

Once, after the usual exchange, I asked: "What's Nick coming in for?" A short pause ensued. I think his father needed to catch his breath. Then he said: "For a revision of his heart. A valve is too narrow." I felt his anxiety. This was the third time surgeons would enter now ten-year-old Nick's chest, and with each procedure the risks rose. Nick was only halfway to adulthood, and he faced multiple surgical procedures in his future.

Approaching a medical procedure that requires anesthesia is stressful for everybody: the patient, the family, the surgeon, and me. The day I no longer feel stress when caring for a newborn is the day I should search for a new career. Babies' conditions change in an instant, with little to no time to make a treatment decision. The old truism is this: *It's hard to kill a baby, but it sure is easy to hurt one.* The TTI (time to injury) from the start of wavering vital signs to lasting harm is a fraction of what it takes to harm an adult. Anesthetizing babies requires extraordinary vigilance.

For the patient, anxiety ends with the loss of consciousness, a magical and momentary transformation. But for those left behind, the anxiety is unrelenting. Waiting for the time of reunion slows the clock, with seconds seeming like minutes, minutes like hours.

AS I ENTERED THE pre-anesthesia holding room, Nick sat up on his gurney, and for the tiniest fraction of a second a flash of light bounced from his chest. In this small, dark room in the interior of the hospital, lacking any natural light, the cart occupied most of the space, with only a couple of feet on the far side for his parents to stand and an equal amount on my side. The only light in the room streamed through a door and reflected off of Nick's chest.

Cheerful as always, Nick's parents greeted me with familiar smiles, even on this day of heightened anxiety.

With time of little concern, I slid a chair onto my side of the gurney. Sitting rather than standing creates the illusion that a physician spends more time with a patient. For me, sitting means more. It puts me at eye level with my patient, rather than peering down from above, and allows me to speak *with*, not to, the child. I hope it's taken as I intend it: as an indication of genuine caring and assurance.

As we spoke, I noticed that the glare from Nick's chest came from a gold Saint Christopher medal—the Catholic patron saint for travelers. The medal showed Saint Christopher carrying a baby—the baby Jesus—on his shoulder with a caption intoning protection. Surgery is a journey for all involved, and the faithful seek divine intervention and invoke a guardian for spiritual comfort. At the end of our discussion I turned to his parents.

"Would you like Saint Christopher to stay with Nick through the procedure?"

"You don't have to," his mom responded.

I persisted. "It's no trouble." This was my attempt at lessening his parents' anxiety.

"Thank you."

With the Saint Christopher medal in hand, Nick and I headed for the OR. When we got there, I pinned the medal to the sheet covering the OR table, next to Nick's head.

Many hours later, the procedure very long because of scarring from previous surgeries, Nick was transferred to the intensive care unit. I was upbeat talking with his parents because everything had gone so well. I went to my office to complete some of the paperwork that medical school doesn't warn you about, fulfilling all the bureaucratic requirements.

My pager went off; the intensive care unit was looking for me. Thoughts of complications raced through my mind. I ran through the list of all the possible problems and how to treat them as I rushed to a phone.

"It's Doctor Jay. Someone paged me. Probably about Nick."

"It was me, Doctor," the nurse caring for Nick said. "I was calling because the parents asked if you knew what happened to the Saint Christopher medal."

"Oh, shit!"

My pulse pounded worse than if there had been a complication.

"Excuse me?"

"I'll get back to you."

Before the receiver settled onto its cradle I was running through the halls, back to the operating room. It had been fully cleaned, the linens and trash removed. I searched the shelves, carts, and drawers to no avail. I asked a nurse who was stocking the room's supplies and hadn't been part of the case if she had any idea where the medal might be.

"I didn't see any medal, Doctor. I just got here."

The nurses from the case had left, so there was no help. My heart sank.

"What happens to the linens after the case?"

"They go to the soiled-linen utility room. It's right behind the anesthesia supply room."

I had passed this room thousands of times before, unaware of its purpose. I hoped to get there before the laundry was removed. Breathless when I arrived, I opened the door and my eyes landed on a mountain of red. The dirty laundry from each case was placed in a large red plastic bag and brought here. The back wall of the room, about twenty feet in length, was covered with dozens of these bags containing the sheets, drapes, towels, and blankets from all the day's cases. They formed a mas-

sive pile at least five feet high. Thinking those from the
most recent cases were likely on top, I began pulling the
bags down and searching them one by one. No Saint
Christopher. After ten bags or so, I gave up.

With my head hanging, I dragged myself to the ICU.

I mustered the courage to look at Nick's parents.
"I'm sorry. I looked everywhere," I said in a voice
reserved for relaying horrible news.

"Don't worry, Doctor Jay," Nick's mother said. "It
wasn't that important."

She appeared sadder for me and my bumbling ways
than for her own loss.

"It was, and this is just inexcusable."

"No, it's not. Really, it isn't that big a deal."

Despite being proud of the medical care I had provided, despite Nick's outcome, I left defeated.

"What's wrong?" my wife asked at dinner that night.
I couldn't hide my disappointment.

"I lost God today."

"What?"

"I lost a child's Saint Christopher medal."

I relived the events of the day and felt miserable.

Nick recovered well and went home in less than a week.

I LANGUISHED OVER THAT medal. I felt the pain as
if that Saint Christopher medal was pinned through the

skin on my chest, a constant reminder of my failing. The concept of no good deed going unpunished came to mind. It's easier not to take objects of value, objects of comfort for either the child or the parents—from blankies to stuffed animals to religious icons—back to the procedure area with the kids. Then nothing can be lost. But that's defeatist.

All the books and articles ever written about children and anesthesia teach how to provide safe pediatric anesthesia care, but none teach how to care for the kids and their parents. I was determined to hold fast to my belief that any amulet, talisman, prayer book, or other object of faith that might ease a family's anxiety could be taken into the OR and remain near their loved one's head throughout. I committed myself to making sure no patient ever lost such a talisman again while under my care. I have lost nothing since.

ABOUT SIX MONTHS LATER, a message taped to my office door listed a familiar return number.

"Hi, Doctor Jay. Are you available . . . ?" Nick's father, always his father, asked.

"Absolutely."

He was as gracious as ever.

"Well, I am happy you still want me."

"Why would we change?"

The case was a small follow-up procedure, and when the day arrived, the surgery was uneventful and Nick was soon in the recovery room.

About a half an hour later, the recovery room paged me. The nurse was asking me to go see his parents again. When I arrived in the recovery room, they were beaming.

Nick's father looked at me and asked: "Doctor Jay, we were wondering if you knew where this Saint Christopher medal on Nick's bed came from?" He was holding the medal.

I was stunned. A smile grew on my face.

"God's work is mysterious," I said.

As I walked away, the pain of that lost Saint Christopher medal piercing my skin evaporated.

FIVE YEARS PASSED, and one day I received a call from our administrative assistant, asking if I could stop by the office. There stood Nick and his parents. They were in town and had stopped by to see me. Nick looked wonderful. He wore a high-necked jersey, so I couldn't see if the medal was around his neck. But his parents hugged me, deeply, and that's what keeps me coming back.

Acknowledgments

———

Thank you to Leslie Rubinkowski, who directed my transition from medical writing ("This is a 16 y/o male who presents with RUQ pain and is to undergo a laparoscopic, possible open cholecystectomy . . .") to writing for a general audience. To Patsy Simms, who accepted me into the Goucher MFA program. To Frank Seleny, Casey Firlit, Andy Roth, and many other colleagues, who have inspired me through the years. To Diana Hume George, Dick Todd, and Suzannah Lessard, who mentored me, and especially to Madeline Blais, who in addition to mentoring me insisted on the title. To Joy Tutela and the David Black Agency, who saw a glimmer of hope in my writing, and to Matt Weiland at W. W. Norton, who polished my thesis into this work. And to the many of thousands of patients who allowed me to enter their lives and provide their care, I am forever grateful.

A Note on Sources

The list that follows is not meant to be exhaustive, but instead represents an overview of the resources I used in researching each chapter.

INTRODUCTION

Anesthesia in the United States 2009. Schaumburg, IL: Anesthesia Quality Institute, 2009.

Bigelow, H. J. "Insensibility during Surgical Operations Produced by Inhalation." *Boston Medical and Surgical Journal* 35, no. 16 (November 1846): 309–17.

CHAPTER 1 | A DEEP SLEEP

Ball, C. M., and R. Westhorpe. "Ether before Anaesthesia." *Anaesthesia and Intensive Care* 24, no. 1 (February 1996): 3.

Brown, E. N., R. Lydic, and N. D. Schiff. "General Anesthesia, Sleep, and Coma." *New England Journal of Medicine* 363, no. 27 (December 30, 2010): 2638–50.

Discovery of Anesthesia by Dr. Horace Wells: Memorial Services at the Fiftieth Anniversary. Philadelphia: Patterson & White, 1900.

"Dr. C. W. Long, the Great Discoverer of Anesthesia." *Atlanta (GA) Constitution*, October 13, 1889, 8.

Duncum, B. M. "Ether Anaesthesia, 1842–1900." *Postgraduate Medical Journal* 22, no. 252 (October 1946): 280–90.

Eckenhoff, J. E. *Anesthesia from Colonial Times.* Philadelphia: J. B. Lippincott, 1966.

Fenster, J. M. *Ether Day.* New York: HarperCollins, 2001.

General-Anesthesia.com. Accessed December 17, 2015. http://www.general-anesthesia.com.

Haridas, R. P. "Horace Wells' Demonstration of Nitrous Oxide in Boston." *Anesthesiology* 119 (November 2013): 1014–22.

Leake, C. D. "Valerius Cordus and the Discovery of Ether." *Isis* 7, no. 1 (1925): 14–25.

Lewis, J. H. "Contribution of an Unknown Negro to Anesthesia." *Journal of the National Medical Association* 23, no. 1 (January 1931): 23–24.

Plomley, F. "Operations upon the Eye." *Lancet* 48, no. 1222 (January 1847): 134–35.

Roland, C. G. "Thoughts about Medical Writing

XXXV. 'Let's Call It Hebetization.'" *Anesthesia and Analgesia* 55, no. 3 (May 1976): 366.

"The Wacky History of Nitrous Oxide: It's No Laughing Matter." *DOCS Education* (blog). July 20, 2015. http://docseducation.com.

Wood Library-Museum of Anesthesiology. "History of Anesthesia—Interactive Timeline." Accessed December 06, 2016. https://www.woodlibrarymuseum.org.

CHAPTER 2 | COMMAND CENTER

Cooper, J. B., R. S. Newbower, C. D. Long, and B. McPeek. "Preventable Anesthesia Mishaps: A Study of Human Factors." *Anesthesiology* 49, no. 6 (December 1978): 399–406.

Lohr, K. N., and H. B. Brook. *Quality Assurance in Medicine: Experience in the Public Sector.* Santa Monica, CA: Rand Corporation, 1984.

Newbower, R. S., J. B. Cooper, and C. D. Long. "Learning from Anesthesia Mishaps: Analysis of Critical Incidents in Anesthesia Helps Reduce Patient Risk." *QRB. Quality Review Bulletin* 7, no. 3 (March 1981): 10–16.

QFD Institute. "Deming Influence on Post-war Japanese Quality Deployment." Accessed August 22, 2013. http://www.qfdi.org.

US Department of Defense. "DoD News Briefing—

Secretary Rumsfeld and Gen. Myers." February 12, 2002. http://archive.defense.gov/Transcripts/Transcript .aspx?TranscriptID=2636.

CHAPTER 3 | THE FIVE A'S

Burney, F., and P. Sabor. *Journals and Letters of Frances Burney*. New York: Penguin Classics, 2001.

Cooper, A. "An Anxious History of Valium." *Wall Street Journal*, November 15, 2013.

Irving, J. "Trephination." In *Ancient History Encyclopedia*. May 1, 2013. http://www.ancient.eu.

Kane, J. "Historical Ties to Rubber Gloves, Beverages, Freud's Nightmares." *The Rundown* (blog). *PBS NewsHour*. October 17, 2011. http://www.pbs.org/news hour/rundown.

Lathan, S. R. "Caroline Hampton Halsted: The First to Use Rubber Gloves in the Operating Room." *Proceedings (Baylor University Medical Center)* 23, no. 4 (October 2010): 389–92.

Malignant Hyperthermia Association of the United States. Accessed December 1, 2016. http://www.mhaus.org.

Meyer, F. "Mrs. Winslow's Soothing Syrup—Oooh So Soothing." *Peachridge Glass* (blog). January 5, 2013. http://www.peachridgeglass.com.

"'Mother's Little Helper,'—Valium." *US History* (blog). May 5, 2010. http://sadieushistory.blogspot.com.

"Mrs. Winslow's Soothing Syrup for Children Teething.; Letter from a Mother in Lowell, Mass. a Down-town Merchant." *New York Times*, December 1, 1860.

National Alliance of Advocates for Buprenorphine Treatment. "A History of Opiate Opioid Laws in the United States." October 29, 2013. http://www.naabt.org.

Raghavendra, T. "Neuromuscular Blocking Drugs: Discovery and Development." *Journal of the Royal Society of Medicine* 95, no. 7 (July 2002): 363–67.

CHAPTER 4 | RAILROAD TRACKS

Barsoum, N., and C. Kleeman. "Now and Then, the History of Parenteral Fluid Administration." *American Journal of Nephrology* 22 (2002): 284–89.

Cannard, T. H., R. D. Dripps, J. Helwig, and H. F. Zinsser. "The Electrocardiogram during Anesthesia and Surgery." *Anesthesiology* 21, no. 2 (March–April 1960): 194–202.

ECG Library. "A (Not So) Brief History of Electrocardiography." December 4, 1996. http://www.ecglibrary.com.

Kutz, S., and J. P. O'Leary. "Harvey Cushing: A Historical Vignette." *American Surgeon* 66, no. 8 (August 2000): 801–3.

Mallon, W. J. "E. Amory Codman, Surgeon of the 1990s." *Journal of Shoulder and Elbow Surgery* 8, no. 2 (March–April 1999): 204.

Zeitlin, G. L. "History of Anesthesia Records." APSF [Anesthesia Patient Safety Foundation] 25th anniversary edition. Wood Library-Museum of Anesthesiology. Accessed June 27, 2016. www.woodlibrarymuseum.org/news/pdf/Zeitlin.pdf.

CHAPTER 5 | FEAR OF THE MASK

"Hippocratic Oath." In *Encyclopaedia Britannica*. Accessed March 1, 2013. http://www.britannica.com.

"History of Medicine." In *Encyclopaedia Britannica*. Accessed December 17, 2015. http://www.britannica.com.

Poland's Syndrome. "Frequently Asked Questions about Poland's Syndrome." Accessed January 15, 2013. http://www.polands-syndrome.com. (Site no longer active.)

White, W. D., D. J. Pearce, and J. Norman. "Postoperative Analgesia: A Comparison of Intravenous On-Demand Fentanyl with Epidural Bupivacaine." *British Medical Journal* 2 (1979): 166–67.

CHAPTER 6 | NOTHING BY MOUTH

Knight, P. R., and D. R. Bacon. "An Unexplained Death: Hannah Greener and Chloroform." *Anesthesiology* 96 (2002): 1250–53.

Mendelson, C. L. "The Aspiration of the Stomach Contents into the Lungs during Obstetric Anesthesia." *American Journal of Obstetrics and Gynecology* 52 (August 1946): 191–205.

Rusch, D., L. H. J. Eberhart, J. Wallenborn, and P. Kranke. "Nausea and Vomiting after Surgery under General Anesthesia." *Deutsches Ärzteblatt International* 107, no. 42 (October 2010): 733–41.

"The Short, Sad Life and Tragic Death of Hannah Greener." *Brian Pears* (blog). May 30, 2015. http://www.bpears.org.uk.

CHAPTER 7 | HEARTBEATS

Conger, K. "For Transplant Patients, the Teenage Years Are the Most Precarious." *Stanford Medicine Magazine,* Fall 2007.

Krishnamurthy, V., C. Freier Randall, and R. Chinnock. "Psychosocial Implications during Adolescence for Infant Heart Transplant Recipients." *Current Cardiology Reviews* 7, no. 2 (May 2011): 123–34.

CHAPTER 8 | A MOST UNUSUAL PATIENT

Eger, E. I., L. J. Saidman, and B. Brandstater. "Minimum Alveolar Anesthetic Concentration: A Standard of Anesthetic Potency." *Anesthesiology* 26, no. 6 (1965): 756–63.

Fish, R., P. J. Dannerman, M. Brown, and A. Karas. *Anesthesia and Analgesia in Laboratory Animals.* Salt Lake City, UT: Academic Press, 2008.

CHAPTER 9 | ERRORS EVERLASTING

Jubran, A. "Pulse Oximetry." *Critical Care* 19 (2015): 272.

Pedersen, T., A. Nicholson, K. Hovhannisyan, A. Moller, A. F. Smith, and S. R. Lewis. "Does Monitoring Oxygen Level with a Pulse Oximeter during and after Surgery Improve Patient Outcomes?" *Cochrane*, March 17, 2014. http://www.cochrane .org.

CHAPTER 11 | PAPER CRANES

Aggrawal, A. *Narcotic Drugs.* New Delhi, India: National Book Trust, 1995.

Alvarez, D. J., and P. G. Rockwell. "Trigger Points: Diagnosis and Management." *American Family Physician* 65, no. 4 (February 2002): 653–61.

Centers for Disease Control and Prevention. "Cerebral Palsy (CP)." January 15, 2015. http://www.cdc.gov/ ncbddd/cp.

Centers for Disease Control and Prevention. "Increases